The Whole Story

The Whole Story

Alternative medicine on trial?

Toby Murcott

Macmillan

First published 2005 by
Macmillan
Houndmills, Basingstoke, Hampshire RG21 6XS and
175 Fifth Avenue, New York, N. Y. 10010
Companies and representatives throughout the world

ISBN-13: 978–1–4039–4500–6
ISBN-10: 1–4039–4500–4

This book is printed on paper suitable for recycling and made from fully managed
and sustained forest sources.

A catalogue record for this book is available from the British Library.

A catalog record for this book is available from the Library of Congress.

10	9	8	7	6	5	4	3	2	1
14	13	12	11	10	09	08	07	06	05

Printed and bound in China

Dedicated to the memory of my father,
Ken Murcott (1939–2004)

Contents

acknowledgements

This book needed to feature a wide range of topics and ideas in which I can claim no expertise. Therefore I owe a great many thanks to the specialists who have been so generous with their time and thoughts. When I have got it right, it is they who have made it so. Any remaining mistakes are mine and mine alone.

I have had numerous conversations with many people, either by phone or email, all of whom have contributed in some way – sometimes explicitly as acknowledged in the text, sometimes by keeping me on the right track. For these conversations I would like to thank Bernadette Carter, Iain Chalmers, Sabine Clark, Rachel Clarke, Mike Cummings, Robert Dingwall, Edzard Ernst, Jane Gallagher, Jenny Gordon, Gill Hudson, Janice Kiecolt-Glaser, Kate Kuhn, Catherine Law, Richard Nahin, Robin Lovell-Badge, Alan Marsh, Hannah Mackay, Andrew Moore, Stephen Myers, Ton Nicolai, David Peters, Paolo Roberti, Virginia Sanders, Aslak Steinsbekk, Andrew Vickers, Harald Walach and Graham Ward. I would like to extend an additional thank you to John Hughes for permission to describe his as yet unpublished data and to Paul Drew, Louise Fletcher and Virginia Olesen for additional help. I am also very grateful to Christine Barry, John Chatwin and Zelda Di Blasi for permission to draw on material from their PhD theses. I also owe a special debt of gratitude to Hilly Janes for allowing me the opportunity to explore some of these ideas in print over the last year.

There are two people who have been a particular help to me both by being generous with their ideas and with their contacts. Paul Dieppe and George Lewith, in different ways, gave me keys to the world of complementary medicine research.

The support and advice of my editor, Sara Abdulla, as well as her efforts on the manuscript, have been invaluable. It's a testament to her courage, or perhaps foolhardiness, that she was prepared to take on a novice author, and she has earned my considerable gratitude as a result.

The process of writing the book would have been impossible without the tolerance of my friends and family who have accepted my constant refrain of 'when the book is finished' with great good humour. Two people in particular have helped me far more than I can possibly express. Anne Murcott patiently guided me through a crash course in sociology and put her academic experience at my disposal: as she insists on putting it, it has been quite handy that she is also my mother. Above all, I am especially grateful to Kerry Chester, my partner and companion, for generously allowing herself to take second place to my writing and for putting up with a one-track monster in the house.

preface

I spent eight years as a research biochemist. Throughout that time I lived with Sam, a large, ginger neutered tom cat. He had moved in with no fur on his belly and back legs – apparently as a result of fleas – and a medicine cabinet of powerful and expensive steroids. I decided that if fleas were causing Sam's baldness then I would deal with them rather than spend my meagre funds on cat steroids. So I bought him a flea collar and dusted him down with flea powder. Before long his fur had grown back.

Ten years later Sam became lethargic and lost interest in his food. The vet diagnosed kidney failure and gave him a few weeks, or at most a couple of months, to live. I'd grown very fond of Sam, so despite it being against my better, scientifically trained, judgement, I followed a friend's advice and took him to a homeopathic vet.

Slightly to my surprise, the homeopathic vet gave the cat a cursory once-over then spent the next twenty minutes grilling *me*. Was he good natured or grumpy? Did he like warm spots to sleep in? How did he eat his food? In the end he announced that Sam was an angry cat and that, as it often did in cats, his anger had settled on his kidneys – and by the way had he ever lost his fur? When I said that he had, the vet replied that it was a common reaction in angry cats. I kept my thoughts to myself, knowing that the best explanation for his fur dropping out was a flea allergy and that removing the fleas cured the problem. Angry kidneys had not till then figured anywhere in my thinking, let alone my biochemical training. But having got this far, and confident that the pills would do no harm – I assumed them to be sugar pills – I duly gave Sam the homeopathic remedies prescribed.

The vet's parting shot had been to warn me that his fur might drop out again and not to worry if it did. That would be the remedy drawing out the anger from his kidneys and making him better. A week later we started to notice clumps of ginger fur on the carpet, and sure enough, the baldness had returned in exactly the same pattern as the earlier fur loss. It soon grew back and Sam went on to live for another year, beating the conventional vet's prognosis by a considerable margin.

Was this my damascene moment? Did I convert to the faith of homeopathy and abandon my scientific career? Not a bit of it. My scepticism remained – remains – intact, yet my scientific training had taught me not to dismiss uncomfortable observations out of hand. It started me thinking: how could it be explained?

There are at least four possible explanations for what happened to Sam: therapeutic effect; coincidence; placebo; and conjuring trick. Working through each alternative rapidly became less interesting than considering the grounds for deciding which was the most reliable explanation. Rather than ask whether or not such therapies work, the first questions are:

How can we tell if they work? What methods do we have that will tell us? This book is the result of pursuing these questions.

Toby Murcott
Bristol, Spring 2004

1

introduction

The global spend on alternative medicines is $60 billion a year and rising. In France, 75% of the population has used some form of what is often also called complementary medicine. That figure is around 50% in the UK, 42% in Canada, and 35% in Norway. More than three-quarters of German pain clinics offer acupuncture. Australians spend A$2.8 billion and the Europe-wide market for herbal remedies is over €600 million and growing, while Americans spend as much as $47 billion each year on what they know as alternative therapies. Complementary medicine, alternative medicine – call it what you will – unorthodox treatments are now the fastest growing sector of many health care systems.

Something is happening to the way we think about our health. Not quite a revolution, more of a sea change, a shift away from our being passive recipients of doctors' wisdom towards becoming active participants in our own health care. Patients arrive at the doctor's surgery armed with their own views on how their bodies work and what can be done to heal them.

Walk into a health food store in Dunedin, the southernmost city in the world, right now, and you can read the latest (winter sport 2004 as I write), edition of *Health and Herbal News*. It takes issue with the idea that prescription medicines are safe and duly approved by the New Zealand Government: 'regretfully the truth is far removed from perception'. Flicking through you might spot the article about the way 'most prescription drugs don't work', or the one about how you can 'ease stomach discomfort with slippery elm'. A thorough read reveals a section called 'Research Review' complete with reports of 'scientific studies (which) prove garlic's effectiveness'. Anyone across the Pacific leafing through the Manhattan Yellow Pages or those of Oakland, California, to the category Physicians and Surgeons, will find listed entries for Acupuncture, Alternative Medicine, Chiropractic, Holistic Health, Homeopathy, Naturopathic and

Osteopathic Physicians alongside those of Pediatrics, Hematology and Gynecology. On the other side of the Atlantic, a leaflet pushed through the letter boxes of Islington in north London advertises a newly opened suite of therapy rooms – 'a stunning holistic centre' offering an 'exceptional and diverse range of complementary therapies' including Craniosacral therapy, Energy healing, Metamorphic technique, Reflexology and Reiki.

This is probably to be expected in the rarefied districts of Islington or Manhattan, home to well-heeled baby-boomers. Dunedin, though, is a city of more modest means. Perhaps even more striking is 'Dr & Herbs', a small shop selling Chinese remedies and offering acupuncture in Bluewater, a new, and vast shopping mall south east of London. Bluewater expressly caters for a *mass* market. In the UK complementary medicine is now a key retail commodity. Boots – one of the best known drugstore chains, with operations in 130 countries – began selling herbal preparations and aromatherapy oils in 1991. In December 2002 one of the major UK supermarkets, Tesco, was reported to have bought a majority share in a prestigious London complementary medicine clinic (established 1987) that was, incidentally, opened by the Prince of Wales.

Alternative or complementary medicines and therapies have become a branch of health care. Driven by consumer demand, only marginally regulated and offering therapies that many scientists reject as absurd, these 'treatments' are mounting a challenge – not easily ignored – to several major aspects of medical care, from means of delivery to modes of action.

Many of the treatments bundled together under the heading of 'alternative' are far older than the conventional medicine they are supposed to complement. Acupuncture dates back thousands of years; likewise massage and reflexology. Homeopathy began at the end of the 18th century, long before antibiotics and heart transplants. Herbalism is perhaps the most ancient of all and certainly pre-dates the evolution of humans.

Our close relatives, chimpanzees and gorillas, eat several medicinal plants and seem to have an understanding of which diseases they alleviate.

Modern medicine is the new kid on the block, and a very successful one at that. For a while it looked like what we now call conventional medicine had swept away all before it, at least in the developed world. In just the last half century, antibiotics, vaccines and surgery have saved countless lives and transformed innumerable others. And yet complementary therapies are staging an unstoppable comeback.

Behind the scenes is a tussle. On one side are those proclaiming the virtues of complementary therapies; on the other, those deriding them as unproven, potentially harmful, nonsense. To complicate things further there is a comparatively recent addition to the fray: integrated or integrative medicine that attempts to merge the best of both worlds.

In the midst of all this are claims and counter-claims about what kinds of therapy do or do not work. One faction wants to place the body's own ability to heal itself centre stage. Another feels that therapies should be independent of state of mind. Another argument is between those who reject many complementary therapies on the basis that they are totally unscientific and those who argue that they might be using as yet unexplained mechanisms of action. Yet another is between those who want to put individualized care at the heart of medicine and those who believe that producing broadly applicable treatments is the way forward. There is even a debate around the question of what does 'work' mean with respect to any treatment?

This thrust and counter-thrust of ideas raises important questions itself. What are these claims based on? How specialized is the underlying thinking? What kinds of science are involved? What methods are being used to justify the claims? These questions are where this book starts, and working out some possible answers are what it is about.

The range of different ideas being brought to bear in the clash between alternative and mainstream medicine is remarkably wide, from immunology and neuroscience through clinical research techniques, pharmacology, sociology, anthropology and a good deal of epidemiology. The names of the sciences involved are comparatively unimportant. What is crucial, though, is the potential for understanding that each discipline and approach offers and the arguments over their relevance, strengths and limitations.

This book does not join in the tussle. Rather it stands on a hilltop overlooking the arena trying to see and report back on who is grappling with whom and how, and (tentatively) what might be making headway. This book is not going to answer the question 'does acupuncture work for back pain?'. It will, though, shed light on why we do not yet have any good answers to that question.

This is perhaps a more difficult approach, but I hope ultimately a more useful one. There is a saying: 'Give a man a fish and you will feed him for a day. Teach him to fish and you will feed him for life'. The plan is that you'll be better equipped to fish in the swirling waters of complementary medicine by the end of Chapter 10.

○

Complementary therapies have become sufficiently big business for there to be commercial clients interested in analyses of the market. In a report entitled 'Alternative Health care 2003', published by the KeyNote Ltd market research company, some figures highlight just how big the sector is. The market for herbal and homeopathic remedies and for aromatherapy oils only is 'believed to have grown 10% to

15% *per year* throughout much of the 1990s'. While it fell back in 2002 and probably 2003 due to a change in European regulation, the prediction is that it will rise 'to over 6.9% in 2006 and 6.5% by 2007'.

Look, too, at the growth in the number of complementary and alternative therapists practising around the developed world. New Zealand is typical: the New Zealand Charter of Health Practitioners, representing some 8,500 complementary/ alternative practitioners, estimates that there are approximately 10,000 complementary practitioners in a country of fewer than 4 million inhabitants. On the other side of the world, there are more than 31,000 practitioners in the records of the European Committee for Homeopathy, while the UK's Shiatsu Society, formed in 1981 with just a handful of members, now has 1,730. There has been an explosion in the types of therapy available: massage, chiropractic, osteopathy, acupuncture, biofeedback, herbal remedies, homeopathy, radionics, naturopathy, reflexology, spiritual healing, water cures, cupping, iridology, hypnotherapy and more.

Official Australian government statistics reported at least 2.8 million Traditional Chinese Medicine (TCM) consultations (including acupuncture) per year in the country, with an annual turnover of A$84 million. More than 60% of Australians use at least one complementary health care product per year, including vitamin and mineral supplements as well as herbal products, and overall Australians spend about A$2.8 billion per year in the complementary sector – A$800 million on complementary medicines alone. Imports of Chinese herbal medicines to Australia have increased 100% per year since 1993.

The estimates of how many people use complementary medicines around the world vary – in part because data in each country are not collected in the same way and the definitions of complementary or alternative therapy are not consistent. Some sources have 75% of the French using some form of complementary or alternative remedy, whereas other sources say 50%; the percentages for the USA vary from 40% to around 70% and so on.

The same KeyNote report records that across other European countries the proportions taking complementary or alternative medicines vary from 50–60% in The Netherlands to a little less in Switzerland at 40%, with Belgium and Sweden quite close at 30% and 25% respectively; the UK trails with 20%. Various surveys and polls suggest that, broadly speaking, more women than men turn to these therapies. It also appears that the highly educated are of a more complementary bent. Whatever the size of the explosion, the trend is towards including new therapies, rather than replacing old ones. Few people are abandoning orthodox medicine; they are simply using complementary medicine as well.

At the same time, many doctors are embracing complementary medicine. In 2003 the Medical Care Research Unit of the University of Sheffield compiled a report for the UK Government's Department of Health showing that 49% of GPs – family practitioners – offered some sort of alternative treatment, with the majority offering it on site rather than referring to outside practitioners. Some doctors even see these therapies as a way of meeting their government-set targets. In Germany, which has a strong tradition of complementary and orthodox medicines running side by side, many doctors are also homeopaths. Numerous US family practices offer acupuncture, massage, aromatherapy and the like. Research quoted by the Australian Medical Association indicates that nearly half of the GPs included in a survey said they were interested in training in fields such as hypnosis and acupuncture, and over 80% had referred patients for some type of complementary therapy. There are now at least 29 academic journals on the topic and around 50 degree- or diploma-level courses in complementary therapy in the UK alone. In the USA at least 20 higher education institutions offer some form of complementary or integrative medicine courses; there are eight in Australia and upwards of 40 across Europe.

The media, too, have discovered complementary medicine. At one end of the spectrum is the sober, sceptical, view illustrated by a short piece in *The Washington Post* in spring 2004. It reported on an article in the *American Cancer Society Journal* highlighting a range of apparently useless alternative cancer cures and argued for better education for doctors and patients about such claims. In a similar vein, the UK's *The Times* has for over a year had a regular column called 'Junk Medicine', written by its science correspondent. A recent edition pointed out that the vast majority of alternative therapies have not been through the same strict clinical trials as is now required for prescription drugs, and of those that had, most failed to show any significant effect. A slightly different approach is offered by the *Guardian* newspaper, which features a regular column by Edzard Ernst, Professor of Complementary Medicine at the University of Exeter. A doctor by training, Ernst argues that alternative therapies should be carefully and rigorously tested. He is applying the conventions of medical science to what, to some people, are the more nebulous claims of the therapies, and finding some effective but many wanting.

At the other end is the human interest type of media coverage, which at times gives an impression of alternative therapies dealing in miracle cures. *The Times* Saturday Health Supplement, called 'Body and Soul', in which the 'Junk Medicine' column appears, most weeks also features a personal account from someone who had an intractable condition that conventional medicine was unable to treat and found relief only from some form of complementary medicine. I have to declare an interest here as I write a short piece that goes alongside these features examining what, if any, scientific evidence exists to support the treatment. I often have to report that there is simply not enough evidence to be able to draw anything but tentative conclusions.

The real media explosion in complementary coverage has been in magazines.

Gill Hudson, currently Editor of the BBC publication *Radio Times*, has been editing magazines, including *Fitness*, *Company*, *New Woman* and *Eve*, for more than 20 years. Hudson launched the men's lifestyle magazine *Maxim*, now the largest circulation publication of its type in the USA. She traces the rise in interest in complementary and alternative medicines back to the aerobics boom in the 1980s. Health and fitness became something that we could all aspire to and attain, says Hudson, rather than being the privilege of elite athletes. Publications sprang up to cater for this demand and the market began to grow.

Women's magazines started to change too. From their 19th century beginnings they had health pages, but these tended to be written by doctors and conveyed an air of authority, handing down wisdom from upon high. Some two decades ago, editors realized that readers wanted to get involved in their own health care and so started to provide tips for them to do so. Features on alternative therapies began to appear, and gradually treatments that had been considered counter-cultural or just quaint, such as herbalism, aromatherapy or shiatsu, moved to the fore. Today, says Hudson, alternative therapies are an essential element of all women's and lifestyle magazines. In fact, she doesn't quite see why they are called 'alternative' at all, so established are they in mainstream magazine publishing.

Hudson identifies the ageing of the baby boomer generation as one of the key drivers of this change. Now in middle age, this group were young in the 1960s, when authorities of all types were being questioned. While their parents would never have challenged a doctor, no matter what they were prescribed, baby boomers not only question them but go off and seek other advice if they are not satisfied with the answer. Furthermore, they have come to expect to live a good, long life and are not prepared to 'give up' when age starts to take hold.

And then there is the Internet. Even the most cursory web search turns up thousands upon thousands of alternative medicine sites, some clearly well researched and authoritative and some barmy by any criterion. There are also plenty of sites offering advice on conventional medicine, providing considerable detail of near enough the complete gamut of conditions, causes, prognoses and types of remedy. These are the modern equivalent of the sections on 'Diseases, Cure and Prevention of' (*The Home of Today*, published by the Daily Express) or the chapter entitled 'A Medical Dictionary' (*Newnes Everything Within: A Library of Information for the Home*) of popular domestic handbooks of the 1920s and 30s. Add this to the gradual reduction in deferential attitudes towards medicine that became noticeable during the 1970s and 80s and patients are often arriving in doctors' surgeries with lists of questions based on their Internet searches.

'Patients with cancer and other life-threatening conditions often turn to complementary/alternative medicine for a variety of reasons, and a major source of their information is the Internet', wrote cancer specialist Scott Matthews of the University of California in San Diego in the March–April 2003 edition of the journal *Psychosomatics*. In response, Matthews and his team have developed a series of questions to help patients determine the reliability of information on cancer information web sites. The answers to questions such as whether the treatments were for sale online, if the treatment was touted as a 'cancer cure' and if the treatment claimed to have 'no side effects', raise or lower metaphorical red flags – the more flags a web site has, the less reliable its information.

This is a noteworthy attempt to determine the scientific veracity of particular web sites. But what such a questionnaire cannot do is pass comment on the vast amount of patient testimony available. Virtually every complementary and alternative therapy web site, whether attempting to provide dispassionate information or to sell you something, will offer patient testimonials describing the effects of their particular treatment. These have a common theme, which goes something like this: 'I had a condition that was making my life a misery, and the doctors could do little for it. Then I discovered treatment X and I have never looked back'. They can be pretty compelling, particularly to someone suffering from a similar condition.

In her 1980s study of the coverage of medicine in the media, *Doctoring the Media: the Reporting of Health and Medicine* (London: Taylor & Francis, 1999), Anne Karpf noted that mass media treatment of alternative therapy was changing and was no longer as unsympathetic as it had been. Indeed, media support for complementary medicine could be seen as part of an attempt by editors to side with the 'voice of the people' against the domineering medical establishment. Three years later, Clive Seale's study of *Media and Health* (London: Sage, 2002) offers a rather different angle. He argues that the mass media counterbalances its reporting of health scare stories with 'the spectacle of ordinary people displaying exceptional powers when threatened by illness', a genre into which coverage of complementary and alternative treatments readily fits.

○

Complementary medicine has been described as the first patient-led form of health care. It is used most often for chronic health problems like lower back pain, eczema, stress or arthritis, which are not life-threatening but are conditions by which conventional medicine is regularly stumped. More recently, complementary health practices have increasingly been accepted and integrated into palliative care where the aim is not to cure but to comfort those with terminal disease. It seems that doctors in this field are more comfortable with a multidisciplinary approach.

As well as offering succour for intractable conditions, complementary therapies appeal to patients' dissatisfactions with orthodox medicine. In the book *Alternative medicine: Should we swallow it?*, Tiffany Jenkins and her colleagues list a number of the reasons why this might be. They cite disillusionment with: being treated as machines needing to be 'fixed'; reliance on 'artificial' pharmaceuticals with unacceptable side-effects; and short consultations that process people like a factory conveyor belt. By contrast, a session with a complementary therapist will usually last around an hour. The philosophy of these therapies is to empower the patient, giving them an active role in identifying problems and solutions. Most particularly, complementary and alternative medical approaches are typically holistic – concerned to treat the whole person, not simply the specific symptoms of components needing repair.

○

A further explanation about the use of the words *complementary* and *alternative* is necessary. Until about 10 years ago most people giving or receiving the therapies discussed here would have called them 'alternative'. But there has been a move to describe them instead as 'complementary' to emphasize that they are intended to run alongside, rather than in opposition to, orthodox medicine. This description is more common in the UK, Europe and the antipodes than in the USA where 'alternative' therapy is still the most widespread term. Doctors, researchers and many practitioners have overcome this confusion by referring to 'complementary and alternative medicines', abbreviated to 'CAMS'.

The definitions are not clear-cut. The House of Lords Select Committee on Science and Technology's Sixth Report on Complementary and Alternative Medicine reads 'Complementary and Alternative Medicine (CAM) is a title used to refer to a diverse group of health-related therapies and disciplines which are not considered to be a part of mainstream medical care'. The inquiry on which their report is based was set up in the wake of recognition by the UK government that the use of complementary medicine was growing both in the UK and elsewhere across the developed world.

An article in Melbourne's *The Age* newspaper in March 2004 discussed the increase in interest in complementary and alternative medicine in Australia as follows: 'Most doctors would agree that alternative medicine should be approached cautiously. But there is less consensus about "complementary medicine", which the Australian Medical Association describes as embracing acupuncture, chiropractic, osteopathy, naturopathy and meditation – or even less mainstream treatments such as aromatherapy, reflexology, crystal therapy and iridology – used in conjunction with conventional medical treatment'. Here there appears to be a sharper distinction between complementary and alternative than in the House of Lords Report.

Commercial organizations have different agendas and so yet other definitions. The KeyNote report is designed to help investors and businesses, and thus excludes what it describes as 'recreational pursuits', such as yoga and Feng Shui; some types of massage; and systems or disciplines with a religious or spiritual aspect, such as faith healing.

What we have today is a picture – which will no doubt continue to change – wherein the kind of medicine called

'Western', 'scientific', 'conventional', 'orthodox' or 'allopathic' enjoys a distinct advantage, a sort of top-dog status compared with others which are marginalized as 'different', 'unorthodox', 'alternative' and so on. This picture has evolved over a long time. In the UK, for instance, the medical *profession* could be said to have begun with the Medical Act of 1858, which specified what qualifications allowed people to describe themselves as doctors. Critically, the Act distinguished between qualified and unqualified practitioners, but did not stop the latter from practising. A statutory boundary was created to be policed by the General Medical Council, the body set up by the Act. The story was different in detail and dates in the USA, Australia and elsewhere, although it is about the same issues: licensing and (above all) control of who can and who cannot call themselves a doctor.

Even depending on this definition – treatments employed by registered doctors are orthodox and others are complementary – brings problems. For instance, chiropractic is regulated by law in the UK and homeopathy is deeply integrated into the orthodox medical profession in Germany, so do they count as complementary or orthodox therapies?

The late Roy Porter, historian of medicine *par excellence*, took this view: 'In a medical world which is increasingly bureaucratic and technology-driven, the Hippocratic personal touch seems in danger of being lost'. Confidence in the medical profession had been undermined, he posited, driving the renaissance, since the 1960s, of 'irregular medicine', a term some two centuries old.

The eighteenth century was arguably the golden age of 'quackery' – a loaded term, for when speaking of non-orthodox medicine we should not automatically impugn the motives of the irregulars nor deny their healing gifts. Far from being cynical swindlers, many were fanatics about their techniques or nostrums.... From the 1780s the one medicine which would truly relieve gout – it contained colchicum – was a secret remedy: the *Eau médicinale*, marketed by a French army officer, Nicolas Husson, and derided by the medical profession.

(Roy Porter, *Blood and Guts: a short history of medicine*. London: Allen Lane, 2002).

As Porter illustrates, therapies can move from being classed as alternative to orthodox over time and back again.

I have used *complementary* and *alternative* more or less interchangeably throughout this book. Nothing is implied by referring to one technique as alternative and another as complementary – I am merely acknowledging that each is not recognized as part of the pantheon of orthodox medicine. Most of the references in the bibliography refer to CAMS, but I have chosen to use as few acronyms as possible – I don't like reading strings of letters and have no wish to impose them on anyone else!

○

The list of therapies that come under the broad heading of complementary and alternative is large and growing. Likewise, there are a number of ways of classifying this wealth of treatments. They can be divided into physical techniques such as osteopathy or massage; qi (or chi) energy-based such as shiatsu or reflexology; mind-based, such as hypnotherapy or neuro-linguistic programming; or even geomancy, such as crystal healing. The therapies can also be categorized by their origins. Acupuncture and shiatsu are based on the Traditional Chinese

Medicine concept of energy meridians running through the body; psychotherapy has emerged from the western tradition of Freud and Jung. None of these categories is particularly satisfying, as there is frequent crossover of ideas from one to another. This is not unique to alternative therapies: biomedical disciplines are equally fluid and ideas cross from one to another all the time. Genetic factors in heart disease have been uncovered by epidemiology and finessed by geneticists, while cardiologists and general practitioners use the information to treat patients.

The therapies discussed here are used as illustrations. When demonstrating the problems involved in evaluating therapies there is little point talking about those for which very little data exists. Therefore, all the complementary or alternative treatments mentioned in this book have one or both of the following features: they have been subjected to some form of research into their effectiveness or they are being used in significant numbers alongside orthodox doctors in orthodox medical practices. These include chiropractic, osteopathy, acupuncture, homeopathy, Bowen technique, acupuncture, psychotherapy, shiatsu and reflexology. This is a small list compared to the huge, and growing, number available.

Furthermore, this relatively short list implies no judgement either way about the effectiveness of other treatments. The problems of assessing and measuring are just as relevant to aura balancing or bioenergetic stress testing as they are to acupuncture. The absence of a therapy from this book merely reflects the fact that there has been far less, if any, research into that therapy, or that it is rarely included in integrated medicine.

A quick word about how research is done is needed at this point. For the results of a study to be acceptable they have to be published according to a quality control procedure known as peer review. This simply means that before an editor will accept a paper for publication it has to be refereed by other academics with similar expertise. The impact of that research is, in part, dictated by the journal in which it appears. There is an acknowledged pecking order of journals, with the elite typified by ones such as the *Journal of the American Medical Association*, the *British Medical Journal*, the *New England Journal of Medicine* and *The Lancet*. Research featured in these publications is hard to ignore; the corollary of which is that research published in journals further down the pecking order is correspondingly easier to ignore. That said, all the articles quoted in this book are from peer-reviewed journals.

There is one major omission from the list of therapies discussed here: herbal remedies. These are biologically active medicines. They can be tested in more or less the same way as pharmaceutical drugs and their efficacy is as easy, or difficult, to determine. The debate surrounding herbal medicines concerns regulation, safety and conflicts with other prescription drugs that patients might be taking. A significant number of pharmaceuticals available today have their origins in plants, which biomedicine has well established ways of exploiting. While herbal medicines may be classified as alternative or complementary, they are similar for the purposes of testing. It is how to test *dissimilar* therapies that is the theme of this book.

○

Discoveries are made at the limits of scientists' abilities. Physicists push their giant particle accelerators to ever higher energies; biologists delve deeper into the workings of our cells; and astronomers stretch the range of their telescopes to see further across the vastness of space. Complementary and alternative medicines are difficult to study, they require a reach into unknown territory. Like all thriving areas of investigation there are factions, personal animosities and a great deal of passion. There are believers and sceptics, waverers and staunch defenders, advocates and rejectionists – never mind the indifferent. This book is the story of how the latest research into complementary medicine, practitioners and patients is giving medicine itself a thorough examination.

2

medicine's conundrum

I received a package through the post yesterday, a new gadget for my collection of supposedly useful technology. As always I got out my penknife to cut the tape sealing the box. I was tired and nicked my thumb while closing the knife. This was irritating, but hardly life threatening; the plaster I put on it was more to prevent the blood going everywhere than to help the cut heal. True, it might have become infected but that was unlikely and not a real concern.

Had I cut myself 100 years ago I might not have been quite so relaxed. My body's ability to mend itself would have been the same then as it is today, but there was one crucial difference. Were a wound to become infected in 1904, little could be done if my own immune system failed to fight back. A minor cut could kill if it became infected, and often did. Fortunately the past century has seen one of the most significant medical breakthroughs: the discovery and development of antibiotics.

Antibiotics have reduced the threat from infectious disease dramatically. Diseases that used to devastate populations, such as cholera, typhoid and even plague can be tackled if there are enough antibiotics to go around.

The other huge advance is vaccination. The first vaccination, for smallpox, is credited to Edward Jenner in 1796. It was a hundred years before the next one, against rabies, was developed, followed over the next 50 years by vaccines for plague, diphtheria, whooping cough, tuberculosis, tetanus and yellow fever. With a few major exceptions, what antibiotics can't kill, vaccinations can prevent. Virtually everyone in the developed world receives a series of immunizations as a child that saves them from a whole host of potential killers – including measles, polio, tuberculosis, whooping cough and diphtheria. The feather in the vaccinators' cap is the eradication of smallpox. With few pocked faces any more in the West, there are barely even reminders of a disease that used to kill 30% of the people it infected.

Polio is next on the list. The World Health Organization hopes to eradicate it within the next few years. The success of the polio vaccination campaign is most visible, or more accurately, invisible, in the West. People of my father's generation lost schoolfriends to iron lungs after they had picked up the polio virus: it attacks the muscles and can leave victims unable to breathe unaided. John Prestwich is 65 – retirement age in the UK where he lives – and holds the record for the person who has lived longest in an iron lung. Advances in technology have provided him with a portable device rather than an enclosed canister. Hospital wards full of rows and rows of iron lungs have gone forever, and the last few machines are kept for emergencies only. Gone too are the leg braces, limps, wheelchairs and withered limbs that were the most visible reminders of the muscle-destroying infection. Thanks to vaccination, the disease has been wiped out in the developed world. While not yet the end of polio – pockets persist in South Asia and Central and West Africa – this is a significant marker on the way to its eradication.

Until the 1960s it was assumed that scientific advances were largely responsible for the increase in lifespan. Then Thomas McKeown, Professor of Social Medicine at Birmingham University in the UK, suggested that it was improvements in public health and nutrition that had had the bigger impact. McKeown argued that the provision of clean water and proper sewage disposal, the destruction of insanitary slums and the availability of a better diet were responsible for people living longer. There is a very close correlation between the availability of clean water and better sanitation and the reduction in the incidence of water-borne diseases such as typhoid and cholera. Better living conditions have drastically reduced the incidence of diseases like tick-borne typhus, which thrive in crowded housing.

It was a bold claim and appeared to relegate biological science to a bit part in the theatre of human health. Today, though, the

McKeown thesis is seen as missing some important elements, not least because he did not consider the role that doctors might have had in helping people to improve their diet and hygiene. McKeown also ignored the contribution that doctors have made to public health by working to enhance living conditions – sitting on public committees, lobbying politicians and so on.

The debate rumbles on regarding exactly which elements have made the biggest contribution to increasing lifespan: medical interventions, public health or nutrition. Science, though, has contributed to all of these. Biologists lead the way in determining the nature of infectious diseases, discovering, for example, that cholera is the result of infection with a water-borne bacterium and malaria is the result of a parasite passed on by the bite of a mosquito. Without that knowledge these diseases would be far harder to tackle. Malaria, for example, was so named as it was originally thought to be the consequence of breathing bad – 'mal' – air. Closing the windows at night to keep out 'bad air' would have had some success at stopping mosquitoes biting. But it wasn't until British Army doctor Ronald Ross discovered that mosquitoes transmitted malaria that the disease could be fought by preventing the insects from breeding and biting.

Another development that has saved lives is the improvements in emergency medicine. Individuals can now recover from previously fatal traumas. Procedures vary from sophisticated surgery that can reattach damaged limbs to the simple use of pressure on a wound to stop bleeding. This has replaced the old idea of a tourniquet, which was shown to increase the chance of gangrene, which in turn could kill. Even basic first aid training now includes resuscitation techniques that can help people survive a heart attack.

And then there's transplantation. The idea has been around for centuries – replacing worn-out bits of our bodies with parts from another human. Or even from another animal – called xenotransplantation. Xenotransplantation was first tried in the 17th century when bone from a dog was used to repair the injured skull of a Russian nobleman. It is now undergoing a controversial resurgence due to our ability to genetically engineer animals. The hope is that pigs or other animals can be genetically engineered so that their organs, or cells even, resemble human ones, boosting the supply of donor tissues. That is still some way off, and today the only animal organs widely used in human transplantation are pig heart valves.

The number of different organs that can be replaced in humans is extraordinary. Lungs, hearts, kidneys, corneas, livers, pancreases, skin, bone marrow and even entire bones are harvested from living or dead donors and swapped for diseased organs, to extend and improve the recipients' lives. There have even been two attempts to transplant arms and hands, with limited success, and a few doctors are seriously considering transplanting entire faces. We are discovering that human organs can be treated pretty much like car components. You can replace worn-out parts as long as you ensure that the new ones match.

The ingenuity of the surgeons is coupled with that of drug developers. One drug in particular, called cyclosporin, is responsible for more successful transplants then any other. It is an immune suppressant: it tones down the body's natural defences, preventing them from attacking the transplanted organ.

Like an army, the human immune system has a reconnaissance arm that scours the body for invaders. On finding one, it calls in the big guns to destroy the intruder. This is what can happen to a transplanted organ: if the immune system recognizes it as foreign, it will be attacked and killed – rejected, in other words. To prevent this, organs are matched as closely as possible to the recipient; the development of sophisticated matching techniques has greatly improved the success of transplant operations. But a perfect match is possible only between identical twins, so there is always a chance that a recipient will reject a new organ. This is where cyclosporin comes in. It dulls the immune system's senses, allowing a well-matched organ to thrive.

The future of transplantation is even more extraordinary. Alongside xenotransplantation research, the technology is being developed to build organs from our own cells and so avoid rejection. Laboratories across the world are trying to grow artificial organs using many different techniques. Some are persuading cells to take up residence in delicate scaffolds of natural materials such as coral, or are creating artificial ones out of synthetic materials. Others are finding ways to harness our cells' ability to organize themselves into complex organs and tissues. This discipline, called tissue engineering, could provide a way to repair damaged nerves or muscles or even to grow entire new kidneys.

Any whistle-stop tour can only hint at the breadth and sophistication of modern medicine. Diseases that once were fatal have diminished or disappeared; people recover from horrendous accidents; and worn-out bits of the body can be replaced. The major advances in medical science have changed societies and expectations. Birth rates in the developed world have plummeted as it has become the norm for children to survive to adulthood. There are few children left brain-damaged because of measles and few families devastated by the death of half their children from diseases such as cholera and typhoid.

Our confidence in the power of modern medical science to provide a cure is a testament to its success, but it has also left it with a problem. It is powerful, but it is not omnipotent. Many diseases still defeat the ingenuity of every physician, specialist or surgeon. Large and increasing numbers of people are suffering from conditions that are either difficult to treat or incurable. These diseases are principally ones of either prosperity or maturity, so they could only appear in significant numbers once the infectious killers had been wiped out, at least in the developed world.

The World Health Organization's database of death rates and causes stretches back to 1950. Even a quick glance reveals the radical change that has happened over the last half century. The average lifespan has increased by 25 years, and that has brought a change in the illnesses that afflict us. As deaths from infectious diseases have plummeted, those from chronic conditions such as heart disease and cancer have soared. In the USA in 1900 the leading causes of death were pneumonia, tuberculosis, diarrhoea and enteritis, accounting for around 35% of all deaths. By 1999 heart disease killed 32% of the population and cancer 24%, with stroke in third place on 7%. These figures are mirrored across the developed world.

There are over 300 different types of cancer, all with very different symptoms, treatments and prognoses. People with cancer of the breast, uterus or testis have a 75% chance of surviving more than five years, whereas those with liver, stomach or lung cancer have less than a 15% chance of surviving that long. In fact, there are as many different types of cancer as there are types of cells in our bodies. The reason is simple: cancer is the unregulated growth of a single cell. If the out-of-control cell is from the lung, the result is lung cancer; if it's a colon cell then colon cancer ensues. And just as lung cells are very different from colon cells, so lung cancer is very different from colon cancer. Worse still, there are almost as many types of lung and colon cancers as there are types of lung and colon cells.

The wide range of different cancers is just the start of the problem for scientists and doctors. Infectious diseases are caused by invasions, be they of bacteria, viruses or parasites, such as bacterial meningitis, influenza or tapeworm, respectively. Most of these organisms have a physiology that is different from ours, giving drugs something to aim at. Penicillin attacks certain types of bacteria, but fortunately does nothing to human metabolism – apart from those who are allergic to it.

The problem with cancer cells is that by and large they are very similar to normal cells; drugs that will affect them will also harm healthy tissues. This is why many anti-cancer drugs can have such devastating side-effects. It is extremely difficult to find something that will hit just the cancer and leave the rest of the body alone.

The most common types of anti-cancer drug home in on one of the cancer cells' few obvious differences – their faster-than-normal ability to multiply. A drug that targets dividing (proliferating) cells can weaken or wipe out a tumour. But side-effects are manifold. One of the fastest-dividing groups of cells is hair cells. Drugs aimed at rapidly growing cancers also hit these hard, which is why many people go bald during chemotherapy.

Another approach is to identify a weakness key to a particular type of cancer. Prostate cancer, for example, requires testosterone to grow, so drugs that block testosterone release can slow prostate cancer growth. Again there are considerable side effects: this hormone treatment is chemical castration. Men receiving it lose their body hair and can develop breasts and suffer menopause-like hot flushes. And in the end the treatment fails: the cancer becomes insensitive to the hormone and grows regardless, often with fatal results.

Cancer is a local uprising, not an invasion. As any general will confirm, it is far harder to quash insurgents than to fight an alien attack. Autoimmune diseases, another type of 'cellular uprising', pose similar problems. There are forty or fifty different autoimmune diseases, most of which appear to be on the increase and the prevalence of which varies from population to population. An interesting indication of their growth is a document published by Theta Reports, an independent publisher of market reports for the health care sector. Called 'Autoimmune Disease Therapeutics Worldwide', it predicts that the global market for autoimmune disorder treatments is growing by 15% per annum and is expected to reach over $21 billion by the year 2006.

In rheumatoid arthritis the immune system slowly eats away the delicate linings of the joints, resulting in an agonizing grinding of bone on bone. Joints swell, movement becomes painful and patients become increasingly immobile. To stretch the military analogy even further, it is like a rogue battalion attacking an innocent group of civilians. Why this happens is still a mystery.

Painkillers and anti-inflammatory drugs can help. When joints deteriorate too much, a range of artificial ones are available. But there is no cure. Of course, a great deal of research is being done on rheumatoid arthritis and the like, some of which will undoubtedly produce improved treatments with time.

There are many such autoimmune diseases that, like rheumatoid arthritis, have no cure: early onset diabetes, multiple sclerosis, bullous pemphigoid, psoriasis, ulcerative colitis, Grave's disease, pernicious anaemia and more. All are the result of our immune system turning on us and all pose similar problems: preventing disease without hampering our vital protective forces.

Heart diseases are yet another health care headache. While there is a strong genetic component to all forms of heart disease, much of the problem is lifestyle-based. The rise in availability of cheap, fatty, sugary fast foods combined with our increasingly sedentary habits is making obesity the biggest threat to human health of the 21st century.

Drug companies meanwhile are working hard to develop pharmaceuticals that will enable us to eat our fill and then pop a pill to stay slim and healthy. Some are already on the market, such as Xenical®, which prevents fat from being absorbed from the gut, or Meridia®, which works on the brain to suppress hunger. Others are in development, such as Sanofi-Synthélabo's rimonabant or GSK's compound 181771. It costs around one billion US dollars to bring a drug to market, so the investment represented in developing a weight loss pill is enormous. Clearly the financial brains within the pharmaceutical companies believe that it is worth spending the money. In other words, they are gambling that people are going to continue to be obese and that sufficient numbers will turn to chemical rather than lifestyle solutions to the problem.

The real fix to the obesity epidemic is much simpler: eat less, eat better, exercise more. It's a prescription that doctors are handing out worldwide and which most patients seem to be ignoring. The message is coming from everywhere. I act as consultant to a TV programme in the UK that attempts to use shock tactics to drive that message home. Almost everyone knows they need to look after themselves, but it's clear that many people struggle to make the change. I find it hard and I'm lucky: I enjoy vegetables, can drag myself to the gym, can afford decent quality food and know how to cook it. One of the biggest problems is that unhealthy, highly processed fast food is cheap and provides more calories per penny than fresh food. There is no doubt that reducing obesity will cut the numbers of heart attacks and strokes, and there is little doubt that simply telling people what to do is not working.

One of the greatest scourges of a longer-lived population is Alzheimer's disease, the cruel dementia that robs one in 20 people over 60 of their memories, movements and ultimately lives.

Again we know a lot about the molecular, behavioural and physical changes that occur during the progression of Alzheimer's. Yet, at present, there is little doctors can do to halt its progress. As another of the ailments where our bodies turn against us, it is proving extremely hard to treat.

There are more diseases to add to this list, particularly of children, that seem to be rising relentlessly in the affluent West. Take asthma. Statistics vary from country to country, but the published research agrees that the number of asthmatics has at least doubled in the last 20 years. A parallel rise in allergies has caused such concern that the European Union has funded a €29 million investigation, called the Global Allergy and Asthma European Network. Then there is Attention Deficit Hyperactivity Disorder, a behavioural disorder in children that has been linked to diet and allergy. It, too, is being diagnosed with greater frequency. Add irritable bowel syndrome, chronic fatigue and chronic pain, and the range of conditions for which, at present, medicine has little to offer begins to look rather long.

This, then, is medicine's conundrum. AIDS, TB and SARS notwithstanding, medicine has dealt with infectious disease in the West. Almost overnight antibiotics made lethal infections minor irritations and vaccinations took up the slack. These, though, were the easy targets. The diseases that afflict us now are the difficult ones. Chronic conditions, often born from failures in our bodies' own defences, are extremely tough to treat.

Its unlikely that Alzheimer's, rheumatoid arthritis, cancer and the like will ever have a quick, simple, cure analogous to a course of antibiotics. Chronic conditions tend to require long term treatments. Anyone prescribed statins to reduce their cholesterol levels will probably take them for life. Transplant patients have to have regular checkups and many take immune suppressant drugs for ever.

Yet to an extent we have been spoiled by a century of medical success. We expect a ready remedy for heart disease or a pill that will stop the dementia that is eroding our loved one's personality. Meanwhile, how many times has a cure for cancer been trumpeted in the media, and how many of our friends and relatives still die from its many forms? Medicine's success has backed it into an expectation-management corner.

What's more, one of modern science's greatest triumphs is highlighting just how hard it will be to tackle these chronic diseases. The Human Genome Project has identified all the genes that direct how a human develops. It has listed all three billion letters of code that make up our genome and identified around thirty thousand active genes. Many geneticists believe that reading the genome was the easy bit. Understanding it is going to be a much bigger job.

One of the central findings of the Human Genome Project is that there are around one and a half million points in the genome that can differ from person to person. These are called single nucleotide polymorphisms. Everyone has the same number of genes; it's these individual letter changes that make the difference, just as swapping the letter 'i' in 'drink' for a 'u' gives you 'drunk' – a very similar word with a very different meaning.

One corollary of understanding our differences is 'individualized medicine', currently the focus of heavy investment. We know that some people benefit from certain drugs while others find them useless, or even harmful. For example, 30 to 50% of patients with clinical depression do not respond to the drugs prescribed, increasing their chances of dying from a depression-related condition. One reason could be several genetic mutations which affect an enzyme in the liver called CYP2D6 involved in the breakdown of these drugs. A handle on genetic differences means we can start to identify which compounds fit which genetic profiles and so develop drugs tailored to individual needs.

No genetically tailored drugs are on the market as yet, but it is only a matter of time. Iressa®, from pharmaceuticals giant AstraZeneca, is one of the few treatments available for small cell lung cancer. Unfortunately it only works for some patients. Research published in the spring of 2004 by two independent groups has pinpointed a genetic difference between patients that respond well to Iressa and those that do not. It should now be possible to develop a genetic test to determine whether or not the drug should be used for a particular patient. That may not be hugely significant on its own – it only takes two weeks to find out whether Iressa works anyway. The implications, though, are far wider.

Discovering that possession of a particular form of a gene dictates whether a drug will be effective allows researchers to home in on the weaknesses of small cell lung cancer. This opens up the possibility of developing other treatments that can either bypass the genetic element or enhance it. Either way, this extra knowledge, gained from genetic research, should help researchers develop more effective treatments for one of the leading causes of death worldwide.

The impact of genetics will also be felt in chronic conditions. An example has already emerged. The group of drugs known as statins are used to control levels of cholesterol in the bloodstream. High blood cholesterol levels have been linked to chronic heart disease and patients prescribed statins usually have to take them for the rest of their lives. A study published in June 2004 in the *Journal of the American Medical Association* showed that the drugs will be significantly less effective for those with two specific genetic mutations. The authors of the paper write, 'We recognize that these data have considerable pathophysiological interest and provide strong clinical evidence that there may be promise in the concept of "personalized medicine"'.

One potential obstacle to all of this is that the business of drug production is the economics of scale writ large. Pharmaceutical companies make back their billion dollar new-drug spend by marketing it to 10% or more of the world's population. Tailored drugs sound great, but how are they going to be profitable? Are manufacturers going to make up 20,000 versions and ship supplies around the world? Are street corner chemists supposed to have the expertise to mix up the right combinations, or will each doctor have a machine on their desktop? As yet there are no clear answers to these questions.

○

The march of medical science has greatly improved the chances of surviving previously fatal conditions, made hitherto debilitating ones easier to live with and raised our expectations of health care. More and more is being asked of doctors, and one area that has attracted consistent criticism, especially in public sector funded health systems, is the time that doctors can spend with their patients. Figures recorded in 2002 for the average length in minutes of a consultation with a family or general practitioner were: Germany 7.6; Spain 7.8; UK 9.4; Netherlands 10.2; Belgium 15.0; Switzerland 15.6. Figures recorded in 1998 for the USA gave an average of 18.5 minutes. These snatched consultations are deeply dissatisfying for both parties. It is proving difficult to balance the development of hi-tech medicine with the basic, low-tech need of doctors and patients to spend time together.

The situation can be somewhat different with private medical care. Here the doctor is the direct employee of the patient and, broadly speaking, the time that doctor and patient have together is related to the patient's ability to pay. It costs a great deal of money and is not a privilege available to the majority. Medical insurance schemes pay for private health care, but control their spending and can and do impose limits on the amount of access a patient gets to a doctor.

Given the problems that orthodox medicine is encountering with treating the growing number of chronic degenerative diseases it is perhaps not surprising that we are turning in ever greater numbers to alternative therapies. One of the driving forces is the natural instinct to seek a cure. We have come to expect an effective treatment for illness and disease, and if one form of medicine, orthodox, cannot provide it then we will look elsewhere. The other incentive is that complementary therapies offer something more subtle than allopathic medicine. Medics are rarely able to spend the time to help a patient come to terms with living with a chronic condition. Long-term illness can put a great strain on limited resources and there is no 'cured' box to tick at the end – only, to be brutal, a funeral. Modern medical practice does not ignore chronic conditions (far from it), but they are just not its strong suit. Complementary medicine, on the other hand, thrives on prolonged involvement with patients and positively encourages them to return. The financial implications of this should not be ignored. Regularly returning patients provide practitioners with a reliable income and could, in theory, be exploited. It might, though, be considered similar to the old Chinese practice of paying your doctor to keep you healthy rather than restricting your visits to a few expensive ones when you are sick.

Complementary practitioners talk about 'wellness' and 'holistic' treatments, not always offering cures but instead suggesting ways of living with one's lot. They give advice on diet, exercise, pampering, and just generally being nice to yourself. A therapy that engages you and makes you feel part of the process of maintaining your own health is alluring, particularly when the alternative is to take a bunch of pills from a doctor whose whole demeanour says 'I'm stumped'.

It might be natural to assume that allopathic medicine has a lot to learn from complementary and alternative methods where it comes to chronic disease. That complementary medicine welcomes long-term treatments suggests that it is far better at treating difficult and low-level conditions. But many of the assumptions about this type of treatment are just that – assumptions – without any evidence to back them up. There is a wealth – no, an unimaginably large fortune – of anecdotal evidence that complementary approaches help people with chronic diseases to feel better. So too does long-term intensive nursing, which is firmly within the conventional medicine fold. Is there any added benefit offered by complementary and alternative therapies or are they equivalent to nursing care? This question is crucial and extremely difficult to answer. In fact there are probably a whole series of answers, but if any complementary therapies prove effective then allopathic medicine has something to learn; if not, then a billion dollar industry is based on nothing but good old-fashioned tender loving care.

3

the advocates

The most powerful advocates of complementary and alternative medicines are the millions of people who pay for them each year. The figures are extraordinary. Half the population of the UK has visited an alternative practitioner; so have half of Americans, more than half of Australians and three-quarters of the French. Around 3,000 French doctors, 5,000 Polish doctors and 7,000 German doctors are trained homeopaths according to the European Committee for Homeopathy, a Europe-wide association for homeopathy professionals. Users of alternative treatments might pay for alternative treatments once, but unless some need were actually being satisfied it is hard to see why they would continue to spend the vast amounts necessary to explain these figures.

The advocacy of two other groups is crucial to the mushrooming of complementary medicine: the practitioners themselves and the growing number of mainstream doctors who work alongside them. It is perhaps this second group, conventionally trained medical doctors, that is the most interesting. Many of the ideas contained within complementary therapies clash with modern science – ideas such as energy channels in the body, or the ability of an ultra-dilute solution to effect a cure. Yet many doctors, in seeming contradiction to their training, either practice some form of complementary treatment themselves or refer their patients to other therapists.

Doctors tend to be healers first and scientists second. The Hippocratic oath is largely about curing the sick, not understanding why they are ill in the first place. It recognizes science but puts a strong emphasis on humanity and caring. Consider this from a modern version of the oath written in 1964 by Louis Lasagna, Academic Dean of the School of Medicine at Tufts University: 'I will remember that there is art to medicine as well as science, and that warmth, sympathy, and understanding may outweigh the surgeon's knife or the chemist's drug'. It is this healing aspect of medicine that doctors sign up to first and foremost. There are of course those who lean towards research, and for them the desire to understand the diseases they treat is compelling, but for most of the doctors I've talked to, at least, understanding takes second place to helping their patients feel better.

The stereotypical complementary therapist is a rather vague character, possibly long-haired, trailing a whiff of patchouli and having a tendency to tie-dye. This is often far from the truth. Take Dr Mike Cummings, currently Medical Director of the British Medical Acupuncture Society. Mike is straightforward, clear, unsentimental and from a highly orthodox background. He discovered acupuncture in a rather unusual way while part of an institution that is as far from romantic hippiedom as it is possible to get.

As a medical student Cummings was particularly interested in musculo-skeletal conditions – sports injuries, strains, sprains, cramps and the like. Disappointed with the lack of emphasis on these types of common problem, he took himself off on sports medicine courses and spent time in rehabilitation centres. On qualifying he realized that it was not going to be easy to pursue his interests if he followed a conventional medical career. So he joined the Royal Air Force. Here was an organization full of fit, active individuals that gave him ample opportunities to develop his skills in sports medicine. He found plenty of injuries to treat and was enjoying learning how to do so. Then came a surprise. In Her Majesty's armed forces, Mike Cummings encountered acupuncture.

One Wednesday afternoon he was in the medical quarters on camp when he noticed that the door to the senior doctor's office was shut. This was unusual, so he asked the duty sergeant what was going on. 'The Squadron Leader is doing some acupuncture', came the reply. Cummings was stunned. What's more, the sergeant informed him, 'the Queen was paying for it'. The RAF, it turns out, has funds to allow doctors to train in a variety of different practices, and acupuncture is one of them. Armed forces around the world are known for their conservatism rather than their willingness to embrace unorthodox ideas. No wonder Mike was stunned.

Cummings did nothing with this information for a year or so but then found himself running a station medical centre and, his curiosity unsated, decided to take a course in acupuncture. He chose a short, practically based offering run by the British Medical Acupuncture Society; this gave him a grounding in how needles can be used to heal. The conventional medical approach to muscular injuries involves a lot of injections of drugs such as cortisone, and Cummings had used this more conventional form of needle insertion as a mainstay of his work, even though, as he admits, he did not always fully understand how it worked. Until that point, however, he had not considered that the needles themselves might be part of the effect. Trained in pharmacology, he assumed that it was the contents of his injections alone that did the work.

On returning to his practice Mike Cummings began using needles extensively. He found that they could be very useful for diagnosis, an extension of his fingers as he probed painful regions and mapped out the extent of the muscular damage. They also became part of his toolbox for treating the conditions he saw. Tense, bunched muscles could be released by inserting a slim acupuncture needle – an instant relief that he could see working before his eyes. Acupuncture became an important addition to his repertoire of treatments. His evidence was that he could relieve patients' pain quickly and easily. His own observations convinced him that there was something to the ancient idea of sticking needles into people to make them better.

When the time came to leave the Air Force he was unsure how to continue as a musculo-skeletal doctor, as that was how he still thought of himself, rather than as an acupuncturist. Then an unexpected opportunity arose to take over an acupuncture clinic. Naively he expected to turn the practice into a musculo-skeletal clinic and develop a portfolio career in medicine. Instead, the demand for acupuncture was so great that the clinic turned him into an acupuncturist. He now says that had anyone told him when he was in the Royal Air Force that he would become a complementary therapist he would have laughed out loud. Yet this upstanding military doctor has entered the world of complementary and alternative medicine.

Cummings then began to take an interest in the research that had been done on the efficacy and effectiveness of acupuncture. He read the papers, examined the studies and was amazed to find that, according to them, the treatment he used all the time did not work. The published findings were totally at odds with the results he saw in daily practice. Today, after many years involved in research, he reckons that the design of clinical trials is often at fault.

The evidence that convinces the pragmatic Cummings is not an abstract trial where patients are averages and numbers but his own eyes: his patients get better. This personal testimony is a common theme that runs through virtually all that advocates of whatever form of complementary or alternative medicine have to say on the matter, be they doctors, practitioners or patients. First person accounts of success are compelling.

Mike Cummings is one of many thousands of medical doctors adding a complementary therapy to their repertoire. The British Medical Acupuncture Society is a body for doctors and dentists, as well as other health professionals, who also practice acupuncture. But there are physicians who are chiropractors, osteopaths, reflexologists, shiatsu masseurs as well as all those who are homeopaths. In the UK six homeopathic hospitals operate with public funds and medical doctor–homeopaths are common throughout Europe. It is not necessary, though, to train in a particular therapy to encounter it. Many other doctors come into contact with alternative therapies daily.

Kate Kuhn is a General Practitioner, a family physician, based in Buckinghamshire near London. She has a PhD in biochemistry, reads the medical journals and does her best to keep up with the latest thinking. She is also very interested in complementary medicine and takes a broadly pragmatic view. The therapies may not work through any mechanism her scientific background has trained her for, but what benefits her patients is her primary objective.

Like virtually all those I have interviewed, Kuhn believes that complementary therapies have more to offer people with chronic conditions than those in acute emergencies. If a patient has a strangulated hernia or a severed limb, the best response is surgery. However, if someone comes to her with arthritis, for which orthodox medicine can offer very little, she is open to anything that might improve their quality of life.

An increasingly common issue for Kuhn, as with many doctors, is that patients arrive in her surgery asking about many types of complementary medicine. Her solution is to work with the patient to find out what is best for them. She teaches them that one of the most powerful things they can do is keep a diary of how they feel. So, for example, she introduces them to keeping a record of the pain they experience on a scale of 1 to 10. She explains that trying one type of treatment at a time is the best way to evaluate which ones work for them, rather than jumping into three or four different therapies simultaneously. In other words she shows them how to conduct trials on themselves.

In keeping with the Hippocratic Oath's central theme, a major concern for Kuhn is to minimize any potential harm to her patients. And she interprets 'harm' as both physical and economic. Complementary therapies can be expensive – especially in a country where orthodox medicine is available free at the point of use – and if patients receive no benefit from spending their money then they are suffering economic loss for no gain. Kuhn has got to know about many of the complementary practitioners in her area and while many are good she has concerns about a few. She talks with patients who are seeing those therapists and urges them to consider whether they are getting any benefit from the treatment. It is a difficult balancing act and requires thought, dedication and a great deal of commitment. It can also be time-consuming, which is a challenge when the pressure on doctors to see more patients is growing. All the same, Dr Kuhn has found a way of incorporating complementary therapies alongside her practice that works for her and her patients.

Doctors like Kuhn and Cummings make a case for adopting an open-minded approach to complementary and alternative therapies from a western medical point of view. They are persuasive because their advocacy goes against their training. And they are not alone. Legions of others hold similar views. How else to explain the 30,000 European doctors who are trained homeopaths, or the more than 20 universities in the USA offering Integrated Medicine courses, including Harvard Medical School, or the 30% of government-funded health care trusts in the UK offering some form of complementary therapy?

○

On 20 September 1997 the respected medical journal *The Lancet* published a major round up of clinical trials of homeopathy. This type of work, known as a meta-analysis, attempts to combine data from lots of different studies to see what larger conclusions can be reached. Seven researchers, based in Germany and the USA, examined 186 homeopathy experiments and decided that 89 had sufficiently good data, gathered from well-designed clinical trials, to include in their analysis. They reached a surprising conclusion:

> The results of our meta-analysis are not compatible with the hypothesis that the clinical effects of homoeopathy are completely due to placebo. However, we found insufficient evidence from these studies that homoeopathy is clearly efficacious for any single clinical condition. Further research on homoeopathy is warranted provided it is rigorous and systematic.

The language is moderate but the message is clear: the researchers reckoned that homeopathy is more effective than placebo in some instances. They could not say that it was effective for one condition as opposed to another, since the clinical trials they studied were disparate. They called for more research. Would they have made that recommendation had homeopathy been shown to be bunk?

Predictably this caused a bit of a stir. Here was one of the most respected medical journals in the world apparently giving credence to homeopathy. Debate rumbles on about how good an analysis it was, but it has never been retracted.

The body of research in support of homeopathy continues to grow. A more recent literature review, conducted by Dr Robert Mathie, Research Development Advisor of the British Homeopathic Association, appeared in the journal *Homeopathy* in 2003. Mathie found that homeopathy is effective for childhood diarrhoea, fibrositis, hay fever, influenza, pain, side effects of radio- or chemotherapy, sprains and upper respiratory tract infection. He also concluded that homeopathy is unlikely to help headache, stroke or warts.

Good quality research exists for other complementary therapies. A couple more merit a closer look. The first is a study of
acupuncture published in another respected publication, the
British Medical Journal, on 30 June 2001.

This was a single study, not a combination of many trials, of
acupuncture for neck pain. The starting point for the research
is that there was little evidence for the effectiveness of the conventional medical interventions, which included painkillers,
massage, physiotherapy and exercise. As the introduction puts
it, 'Current treatment increasingly includes complementary
methods, of which acupuncture is one of the most common.
There is, however, a lack of evidence to support acupuncture as
an effective treatment for chronic neck pain'.

A total of 177 patients were followed over three years. They
received either massage, acupuncture or a form of sham acupuncture using a laser to give an impression that something
had happened. The study concluded that acupuncture was a
safe and effective treatment for neck pain and that particular
groups of patients seemed to benefit more than others.

Also gaining ground is the Bowen Technique, a form of contact therapy that relies on soft pressure applied to certain key
points. Developed by Australian Tom Bowen in the first half of
the 20th century, it is based loosely on mainstream physiology,
but its philosophy is complementary. Bowen practitioners talk
about 'drawing the body's attention to the problem' and 'encouraging the body to heal itself'.

The Bowen Technique is often offered for 'frozen shoulder' or
adhesive capsulitis. This afflicts roughly 3% of people and usually strikes those over 40, affecting slightly more women than
men. Its medical description is 'a major swelling in the layer of
cartilage that lines the shoulder joint which dramatically
restricts the movement of the joint'. It can be painful and
makes simple everyday actions such as reaching up to a cup-
board or putting on a coat very difficult. It normally clears up of

its own accord in anything from 18 months to four years. Doctors prescribe painkillers, steroid injections and physiotherapy – none of which is particularly effective – and as a last resort, surgery.

An alternative is around five trips to a Bowen therapist. The therapist makes a few gentle rolling movements or soft rubs over the area with his or her fingers, or more usually thumbs. It is far gentler than most massage and totally non-invasive.

Bernadette Carter is Professor of Children's Nursing at the University of Central Lancashire. She had a persistent soreness in her ankle and was referred for Bowen therapy by her GP. She was very surprised that such a mild intervention helped and decided to investigate further. She discovered that Bowen was also popular for frozen shoulder, and decided to do some research of her own.

Professor Carter recruited 20 patients with frozen shoulder and started by measuring the mobility they had in their affected joints and asking them how much pain they experienced. Her team measured active mobility by asking patients to move their arms and looked at passive mobility by themselves gently moving patients' arms. Patients also completed the McGill Pain Questionnaire (a commonly used method of assessing pain), grading how much pain they experienced in their affected shoulder on a scale of 1 to 10, where 1 is mild and 10 is unbearable. These are well-established techniques. Each patient was then given up to five sessions of Bowen Technique frozen shoulder moves. This was not a randomized trial – it was a pilot study to see whether or not further investigation was warranted. The control for the experiment was that patients' frozen shoulders were compared with their unimpaired sides.

All showed some improvement, Carter's group concluded. Seventy per cent regained normal mobility by the end of the study – an improvement that persisted after the last Bowen treatment. Professor Carter published her pilot study in the journal *Complementary Therapies in Medicine* in December 2001. The only conclusion she offered was that Bowen Technique for frozen shoulder warranted further research.

'Warrants further research' is common to the early development of most medical treatments. If funding is forthcoming, larger, more controlled studies evaluate the treatment more carefully. The Bowen Technique has taken one of the essential first steps towards to becoming an evidence-based treatment for frozen shoulder.

○

The first time many people come across complementary therapy is in the shape of a story. 'I was desperate and didn't know what to do and then I discovered...'. The first person account is the lifeblood of women's magazines – pick any one at random from the news stand and the chances are it will have at least one tale of triumph over the odds. The column I write for *The Times* newspaper gives a scientific perspective on just this kind of story in a series entitled 'It Worked For Me'.

Testimonials to different treatments are perhaps most prevalent on the Internet. Just try typing the name of a particular therapy plus 'testimonial' into a search engine and you will be inundated with stuff like: 'I arrived at the clinic barely able to walk... I was very skeptical... as I lay on the treatment table after Mark had administered a few seemingly innocuous touches, I distinctly felt the pain start to move down the side of my body... I was able to walk to the station and take a train home'. Or 'If you haven't tried Reflexology and Holistic Medicine I'd highly recommend it. I know it's been a godsend for me'. Or 'I feel a huge difference (surprisingly to me)... it's funny... I feel more understanding or not so quick with a temper... AND I dont get tired easily at night... normally I would feel lazy and tired when it gets dark outside... but I've been going to bed late and waking up at a decent hour and I feel great...'.

Storytelling is probably as old as humanity itself: from Greek myths to tales of the court of King Arthur, from Grimm's fairytales to the legends of the Ik people of East Africa. Narrative is a form that holds our attention. And not simply because of the sudden twist or the cautionary tale, the dose of uplifting heroism or the wondrous, almost miraculous, dénouement. More than that, narrative enables us to share, to learn, to reflect and to empathize.

Small wonder then that stories can often be the starting point for discovery. It was anecdotal evidence that individuals who contracted cowpox did not catch smallpox that encouraged Edward Jenner to experiment with cowpox vaccinations. It was anecdotal evidence that led Dr John Snow to discover the cause of an outbreak of cholera in London in 1854. He followed up stories of who had contracted the disease and who had not and worked out that all those infected had drunk from the same pump. As soon as the pump was disabled, the epidemic died out. The combination of anecdotal evidence and Snow's hunches about the disease undoubtedly saved many lives.

○

Hannah Mackay is a shiatsu practitioner with a PhD in psychotherapy and has been researching shiatsu for a number of years. We talked primarily about her research, but at the end of our conversation I asked her why she practises and what she gets out of it. She told me of her success with neck and shoulder pain, something she feels skilled at and can get good results for. She told me of some of her ability to induce labour in overdue pregnant women. Then she became a little more thoughtful and moved off in a different direction. Instead of listing a number of benefits or enjoyments she began to tell me about one of her clients, 'X'.

X arrived at Mackay's practice complaining of back pain and sciatica – pains in the legs that can be extremely severe. Hannah began to treat X and almost immediately became aware that her client was 'very closed down'. Her diagnosis was that her client's body was unable to heal itself because it was 'stuck'. So she began to work on the patient's back and legs to 'open up' the body and allow it to move more freely. This did produce some reduction in the pain quite quickly, so when X returned for a second session she continued in the same vein. Once again the client reported improvement. On the third visit Hannah concentrated her efforts on manipulating meridian points 'to boost the client's energy' in the back and legs. In doing so she says she 'became aware of something frozen deep inside' and had a strong sense it was about X being bullied as a child. She had a feeling that her client didn't want to put her leg on the floor because she didn't trust it, she didn't feel safe. At the end of the session she told the client her suspicions. The client responded by telling Hannah about a trip they were due to take with their father who had abandoned them as a child. The conversation prompted further recollections, including the realization that the pains had started shortly after their father had made contact after years of absence. According to Mackay, X went away considerably better and had also gained an insight into the concerns they had about going away with their long lost father.

The point here is not the content of the story. It's the way Mackay told it. She repeated herself, hesitated from time to time and seemed to be trying to put into words something that was hard to describe. She referred to the case as a complex one and was trying to explain her experience of talking with her client about their concerns and trying to illustrate the experience of what it was that came to her as she worked on their body. More than once she described the experience as 'interesting' or 'fascinating'. Her recounting of the story reminded me of a young child relating a new discovery or an adult describing a very profound and possibly unexpected experience. She sought to convey that the engagement with her client was vividly real and was experienced on a number of levels. And she seemed to have gained a great deal from the interaction.

This depth of interaction is something that virtually all complementary and alternative therapists emphasize. Its more than just spending an hour with their clients as opposed to the 10 minutes or so available to a harried general practitioner. It is also about engaging and giving of themselves. It's evidently very important and rewarding, and is regularly described as vital to what they do. Many researchers investigating complementary medicine suspect that this deep connection with clients is part of the reason that the therapists get results.

Another noteworthy aspect of some complementary therapies is that they often require practitioners to receive regular treatment themselves, particularly while training. Mackay had to take shiatsu from another therapist in order to become registered with the Shiatsu Society. This is also the case for psychotherapists, most particularly psychoanalysts, and Alexander Technique teachers. It's an interesting requirement for a healer to have to undergo the treatment they dispense. Maybe this routine helps practitioners to maintain a strong empathy with their patients.

Mackay's testimony is also an exemplar of the way complementary therapies link physical ailments – in this case back pain and sciatica – with emotional states.

A central philosophy of Samuel Hahnemann, the inventor of modern homeopathy, was that you treat the patient, not the disease. A homeopathic consultation will include details about the patient's life, habits, likes and dislikes. Is the patient an owl or a lark? Does the patient like spicy or bland food? Do they prefer hot weather to cold, and so on. The homeopath then prescribes a remedy based on the symptoms with which the patient presents as well as on their temperament and personality. This can result in two individuals with apparently identical sets of symptoms being prescribed different remedies. Furthermore, as the condition progresses and the symptoms change, the homeopath may prescribe a new remedy. Some are pretty much universal – arnica is regularly prescribed for shock and bruising – but others are only used for particular personality types. The book *Homeopathic Medicine at Home* by Maesimund Panos and Jane Heimlich lists 19 different remedies for a cold. Aconite, for example, is described as 'better for people who are frightened or restless at night' whereas mercurius is recommended for those who have a 'coated tongue with bad odour from mouth' or are 'very thirsty, even though mouth is moist'.

This concern with treating the patient not the disease is reflected in the more mainstream medical literature as well. An editorial on integrated medicine published in the *British Medical Journal* in January 2001 was subtitled 'Imbues orthodox medicine with the values of complementary medicine'. It said 'Integrated medicine has a larger meaning and mission, its focus being on health and healing rather than disease and treatment. It views patients as whole people with minds and spirits as well as bodies and includes these dimensions into diagnosis and treatment'. It goes on to caution that orthodox medicine has promoted technological solutions to ill health over holism and simple techniques such as changing diet and relaxation exercises. The inclusion of this in such a prestigious journal is not a sign that mainstream medicine is coming around wholeheartedly to the viewpoint of complementary practitioners. Rather, it draws attention to the different approaches of the two camps. The orthodox medical view is that the body is a machine that can be cured by fixing its parts; the complementary or alternative view is that mind and body are linked and must be considered together.

A separation of mind and body in modern medicine is a source of frustration to some of the people who work within both complementary and allopathic fields. David Peters is a conventionally trained doctor, complementary practitioner and Professor of Integrated Healthcare at the School of Integrated Health at the University of Westminster in London. Paul Dieppe is also a conventionally trained doctor, a rheumatologist of international renown and Professor of Health Services Research at the University of Bristol in the West of England. Both talk passionately about the ideas of the philosopher René Descartes.

Between 1628 and 1649, Descartes wrote a series of texts laying out his thesis that the mind and the body are separate entities. They can interact and influence each other, he argued, but the conscious and the mechanical are distinct. These ideas laid the ground for the work of many later philosophers, including Spinoza and Leibnitz. Dieppe laments that the legacy of Descartes' approach is mechanistic medicine. This ignores, he says, the fact that our state of mind affects our health and vice versa. Likewise, Peters respects Descartes' radical thinking but says he is yesterday's man. Science, as he puts it, moves on. Dualism, scientific truth and objectivity have been supplanted by entanglement, uncertainty and relativity. The culture of scientific medicine needs to catch up. In short, complementary medicine majors in two things. The relationship between the patient and the practitioner and treating the entire person and not just a set of symptoms.

Clearly, the bulk of evidence in favour of complementary therapies currently takes the form of subjective, first-hand accounts – though the body of scientifically respectable evidence is growing. The practitioners continue to practice because they judge that their clients benefit. By and large therapists are looking for pragmatic, empirical evidence. The absence of 'hard' data clearly troubles some of them, but it does not drive them from their chosen path. The critics, on the other hand, criticize with the full weight of modern scientific understanding behind them. That body of knowledge is large and powerful and makes a good – but not unassailable – case for the ineffectiveness of many complementary and alternative medicines.

4

the critics

There is so much about complementary or alternative therapies that is open to criticism, from the frankly crackpot ideas
about how the therapies work to the insistence of some therapists that what they do cannot be measured. Our modern
understanding of science demonstrates that many complementary therapies cannot possibly work as described. This
idea, know as prior implausibility, is used by some scientists to
argue that there is little point in even attempting to investigate
them.

Lets start with homeopathy. It is sitting there with a great big
'Kick me!' sign stuck to its back – an invitation too good to
miss. Modern homeopathy was developed towards the end of
the 18th century by Dr Samuel Hahnemann. He based it on the
alchemists' principle that 'like cures like': a small amount of a
substance that causes a condition can also treat it. For example, a tincture of dandelion is supposed to cure bed-wetting.
Dandelion is a diuretic – hence its French name *pis en lit*. So any
child suffering from bed-wetting will be cured, the argument
goes, by taking a small amount of dandelion before they go to
sleep. It's a pretty simple idea and one that has, perhaps, a
smattering of real-world inspiration. Chemicals that can kill are
often used to cure. Eat a handful of foxglove flowers and your
heart could fail, yet taking around one four thousandth of a
gramme a day of digoxin – the drug extracted from foxgloves
– can help keep you alive after a heart attack. In this instance –
not necessarily in general – a large amount kills and a small
amount cures. Not too far from homeopathy, you might say.
The main problem comes with what is considered 'a small
amount'. This is a very major difficulty.

Homeopathic remedies are prepared from 'tinctures'. These
tinctures are made by steeping a dried plant, sea shell, or mag-
nesium sulphate, say, in alcohol for several weeks. Tinctures are
not given to patients. Homeopaths dilute them in a large
amount of water or a water–alcohol mix. They shake the mix-

ture vigorously, then dilute that, then shake and dilute as before. The number of dilutions varies depending on the remedy. A typical preparation might be a series of six dilutions of one part in one hundred, known as '6C' and one thousand billion times more dilute than the original tincture. The resulting liquid or pills – made by spraying sugar tablets with the mixture and allowing them to dry – are the 'remedies'.

The more dilute the remedies the more powerful they are considered to be. So a 6C remedy that is a 1,000,000,000,000 (1 followed by 12 zeros) fold dilution is considered less potent than a 30C remedy which has been diluted 1,000,000,000, 000,000,000,000,000,000,000,000,000,000,000,000,000, 000,000,000 times (1 followed by 60 zeros). Allegedly, further dilutions increase the 'potency' even more.

This is palpable nonsense. A little maths reveals why. One drop of the original tincture contains of the order of 1000,000, 000,000,000,000,000 molecules of any active chemical. After a single 1 to 100 dilution, a 1C dilution, that will come down to around 10,000,000,000,000,000,000 molecules, which is just lopping off two zeros from the end of that very long number above. Each further dilution will remove another two zeros. So by the time the remedy has reached the 6C stage there will be 1,000,000,000 molecules of the original active ingredient left. This is a huge number, but if that active ingredient were aspirin it would represent around a thousand billionth of a gram, a speck so small you'd need an electron microscope to see it.

Remember, though, that 6C is not considered a particularly potent homeopathic remedy. To get it up to high potency, say 30C, many further dilutions occur. At 10 dilutions, 10C, there is just one molecule of the original material remaining. Do that a further 20 times to reach the 30C and there can be no traces left. Many of the highly dilute 'remedies' that homeopaths use have no active ingredients in them. They may be labelled

Pulsatilla Nigricans or *Arnica Montana*, but they are simply water. According to every scrap of scientific knowledge of pharmaceuticals, the only effect that can possibly follow from a homeopathic treatment is the placebo effect. Homeopathic remedies can no more cure kidney disease in a cat than can a glass of tap water.

That's not quite the end of the story. Homeopaths have an answer to this charge: 'water memory'. Shaking is vital, they explain, because it encourages water to 'remember' the shape of the chemicals dissolved in it. It is this memory, they posit, that has the therapeutic effect. Without the shaking, preparations are useless, they report, 'proving' that water has memory.

Water is a fluid. Fluids' atoms or molecules are randomly arranged, unlike the regular patterns in most solids. What's more, fluid molecules are in constant motion. Take two snapshots a split second apart and you get totally different but equally random pictures. Things change, though, when you dissolve something in a fluid: say some table salt, sodium chloride, in water. The water molecules arrange themselves into a sort of cage around the dissolved sodium and chloride ions.

When you remove the dissolved substance this order collapses into disorder – the neat little cage of water molecules rapidly disperses. Without the dissolved chemical to surround, the water molecules become part of the amorphous soup of the fluid again. This is driven by one of the fundamental laws of physics: the Second Law of Thermodynamics. This states that 'the entropy of the Universe increases'. Entropy is the amount of disorder in a system. A stack of bricks is low in entropy; pushing it over increases its entropy. Stacking the bricks again requires effort. Lowering the entropy of a system, in other words, takes energy.

So, to go back to the water cage surrounding the table salt. The cage is formed around the salt molecule and needs that template for it to exist, just as a sumptuous designer dress collapses into a shapeless mess as soon as the wearer takes it off. The entropy of the dress or the water cage increases when either loses its shape and it takes energy to reconstruct them. A water cage will no more retain its shape in the absence of a dissolved molecule than will a dress stand upright in the wardrobe without any support. To do so would contravene the Second Law of Thermodynamics, and no one has ever found any circumstances under which that happens. For something to do so – such as water having memory – would mean the most staggering rewrite of our understanding of, well, everything.

Advocates of homeopathy have one more card up their sleeves: the shaking. The frantic agitation at each round of dilution, they suggest, supplies the energy that keeps the cages of water together, remembering the shape of the molecules they once contained. There's a hitch, of course.

Shaking a solution does impart more energy to it. Indeed you can heat up a container of liquid by shaking it rapidly. It's a slow process: shaking a mug full of water twice a second for a minute would heat it up a mere fraction of a degree. The sides of the vessel hit the atoms or molecules in the liquid, speeding them up. Heat is just a measure of how fast molecules are moving, or more accurately it is a measure of the energy they have due to their movement – their kinetic energy. The molecules in a cup of tea move 10 times faster than those in an ice cold gin and tonic. So when the homeopath shakes a diluted solution it does gain a tiny amount of energy and heats up a fraction of a degree. However, heating a solution in this way would also increase the chance of destroying any ghostly cages of water. As the surrounding water molecules heated up and moved faster they would bash into the cages at higher speeds, knocking them apart.

In short, homeopathy is a dead duck scientifically. It uses remedies that contain no active ingredients and a laughable principle that water has a 'memory'.

Having dispatched homeopathy, let's turn to another commonly used complementary therapy: acupuncture. According to the World Health Organization, in 1990 there were 88,000 acupuncturists in Europe, of whom 62,000 were medical doctors. Over 20 million Europeans used acupuncture. In Belgium 74% of all acupuncture treatments are carried out by doctors; in the Netherlands 47% of the general physicians use acupuncture and 90% of the pain clinics in the United Kingdom and 77% in Germany use acupuncture. In 2002 the British Medical Association surveyed 365 doctors and found that almost half of them had arranged acupuncture treatments for their patients, suggesting that the trend has continued.

Acupuncture practitioners believe that there are meridians throughout the body, channels of energy that influence our organs. When these meridians become blocked, they argue, energy, or qi, cannot flow along them, so we become sick. Inserting needles (in acupuncture) or applying pressure (in shiatsu) at specific points along these meridians, the theory goes, can stimulate the flow of qi and so cure some maladies. It is a system that was developed in China thousands of years ago and has a reputation for improving a number of conditions, including pain, eczema and some gynecological complaints, such as painful periods and pelvic inflammatory disease. There's just one problem: meridians do not, as far as we know, exist.

The human body has been dissected, examined, described and documented in fantastic detail since the time of the Ancient Greeks and Egyptians. Mountains of textbooks describe the pathways of the tiny nerves and vessels that meander beneath our skin. Individual anatomists have spent their entire working lives mapping the nervous system of the head and neck or the minutiae of the reproductive system. The picture they have built up is one based on physical structures than are common to all of us.

Perhaps the ultimate expression of human anatomy is the Visible Human Project run by the National Library of Medicine in the USA. This currently contains almost 7,000 ultra-detailed images of normal male and female cadavers. The bodies have been imaged with X-rays and Magnetic Resonance Imaging (MRI) scanners; then they were sliced into very thin sections and photographed. The male cadaver has been imaged down to a resolution of one millimetre and the female cadaver down to a third of a millimetre. The resulting avalanche of data allows researchers to build up pictures of the human body from many different angles using a combination of different types of imaging. Best of all, the data is available online to all at http://www.nlm.nih.gov/research/visible/visible_human.html. This project allows people to go back again and again and search through the bodies to see if they missed anything last time around. You can only dissect a cadaver once, but this virtual human body can be dissected by a computer program many times.

No dissection or imaging method to date has found traces of meridians, acupuncture points or energy channels running through the body. Anatomists have looked and every time they have found nothing. Meridians, the cornerstone of acupuncture and shiatsu, are just not there.

None of which *proves* that meridians will never be found. It might be that we simply don't know what to look for – perhaps meridians are so subtle or so obvious that we are missing them. There are recent precedents for this in anatomy: for example, in 1998 urology surgeon Helen O'Connell discovered that the human clitoris is far larger than was thought. It is actually quite a substantial organ with arms that surround the opening of the vagina and an extensive structure within the pelvic region. Until that point the clitoris was generally assumed to be just a female analogue of the penis, but considerably smaller. The male-dominated medical profession was not interested in the organ and so had ignored its anatomy. Similarly, it could be that western medicine has until quite recently had little interest in meridians and so hasn't found them yet. One of the problems with this argument is that it is impossible to prove a negative. There cannot be any conclusive proof that meridians do not exist – the closest it is possible to get is repeated failures to find them. Meridian enthusiasts will always be able to say that they might exist.

But the absence of evidence for meridians is compounded by the failure to find any trace of qi energy. A modern doctor in a well-equipped hospital has a raft of techniques with which to peer inside her patients' bodies. X-rays produce detailed snapshots. MRI scanners can produce thorough images of organs as they function. There are meters for detecting nerve signals travelling to and from our brains and ways of examining the electrical signals that our muscles generate. Ultrasound measures blood flow and, of course, monitors the growth of a baby inside the womb. These techniques pick up radio waves, electric currents, sound waves and X-rays. None of them have ever shown any traces of energy travelling along paths – other than our nerves – inside our bodies.

Once again it could be that qi is invisible to all our current technology, or it could be that we are looking in the wrong place at the wrong time or in the wrong way. Once again, absence of evidence is not conclusive evidence of absence.

Take all this together, though, and the obvious conclusion is that the entire thesis of meridian-based therapies is wrong. Whatever happens in an acupuncture clinic is very unlikely to be based on energy moving along meridians within the body.

There is further damning evidence for the non-existence of qi energy and the meridians it travels along – from the practice of reflexology. This is a massage technique that purports to work on the principle that all of the organs and tissues of the body are linked to particular points on the feet. The kidneys are 'connected' to the arch of the foot and the pituitary gland to the centre of the big toe and so on. By massaging the foot practitioners claim to diagnose problems elsewhere in the body and treat them. A kidney problem might be expressed in tenderness in the arch, and careful work on that spot will help improve the condition. Once again these linkages have not been found anatomically, but it gets worse. Different schools of reflexology have different maps of the foot with different linkages. It is reasonable to counter that if the links were real then the map would be universal. All doctors worldwide agree where the organs of the body reside: someone claiming the heart is in the thigh would be treated with derision. Either reflexology is nonsense or the model of different parts of the body being connected to different parts of the foot is nonsense.

○

Osteopathy and chiropractic are two widely used manipulative therapies that are increasingly offered alongside conventional physiotherapy for, among other things, lower back pain, which orthodox medicine struggles to treat successfully. Physiotherapists are happy to accept that osteopaths and chiropractors are skilled manipulators and often work in conjunction with them. Yet once again the basic model of how these treatments work is fundamentally flawed from a scientific point of view.

The creator of osteopathy is generally accepted as being Dr Andrew Still, a surgeon in the Union Army during the American Civil War. He described his discovery as the ability to cure diseases by shaking the body or manipulating the spine. In his autobiography he wrote that he could 'shake a child and stop scarlet fever, croup, diphtheria, and cure whooping cough in three days by a wring of its neck'. The basic principle behind his ideas is that diseases are largely the result of loss of structural integrity of the body; that the bones, particularly those of the spine, are out of alignment. Osteopathy is based on the belief that manipulation of the bones, muscles and tissues can realign the body and so cure disease. Largely rejected by the orthodox doctors of the time, Still set up an osteopathic medical school in 1892.

At around the same time, another American, Daniel David Palmer, was developing another therapy based on the idea that spine misalignment was responsible for medical problems. He proposed a different model from that of osteopathy – that out-of-place vertebrae affect the nerves radiating from the spinal column and cause the illness. Chiropractors are concerned with realigning the vertebrae by manipulating the spine and prompting healing in that way.

Chiropractic and osteopathy both consider medical problems to arise from joints being out of true. Physiotherapists and orthopaedic surgeons agree that the idea that joints are misaligned in the way these therapists describe is untenable. There are just no data from X-rays or body scans that support it. The possibility remains, though, that both osteopathy and chiropractic achieve their success in exactly the same way as conventional physiotherapy, based on modern anatomy and physiology.

○

There are some forms of complementary therapy that rely on what appear to be modern pieces of technology wrapped up in pseudo-scientific claims. There is, for example, a device on sale called the Aqua Detox. This claims to remove toxins from your feet as you rest them in a bath of warm, salty water while an electric current is passed through both the bath and your feet. The proof of this is purported to be that the bath starts off nice and clean and ends up murky brown, with a scum floating on the top. Its mechanism of action prompted a lively discussion in the letters pages of the *New Scientist* magazine in July 2004.

A correspondent reported trying out the device, describing how he watched salt being added to the correct amount and the current switched on. The water duly became brown and murky, but before it became too dark he observed the discolouration coming not from his feet but from the plastic casing surrounding the electrode. Intrigued, he returned to the treatment centre the following week and this time watched the machine in action without putting his feet into the water. It duly went just as brown and murky.

A second correspondent offered a chemical explanation of what was going on. An electric current passing through a solution of salt water will produce a weak solution of sodium hydroxide, and this in turn will react chemically with the oil in the skin to produce what is described as 'a soapy gunge' – a plausible explanation for the scum on the water. The brown colour can easily be explained if the electrodes were made of iron – the brown is simply rust.

Both correspondents offer speculation, not hard evidence. However, there is a well-established principle in science known as Occam's Razor: that, of two explanations on offer, the simpler is more likely to be true. In this case the simple explanation is that a combination of rusty electrodes and oily feet produce the brown scummy water. The more complex explanation is that some as-yet undiscovered mechanism is drawing toxins through the skin of the feet resulting in brown water and a detoxified body. Occam's Razor dictates that rusty electrodes have it.

Out on another equally unscientific limb is crystal healing. This is the principle of using the 'energies' of different crystals to effect 'cures'. Different properties are assigned to different crystals and they are placed on the patient at appropriate points in order to 'energize' the body or 'release energy blockages'. There is perhaps a smattering of scientific inspiration buried within crystal healing. If quartz is squeezed it produces a tiny electrical current – the piezo-electric effect. The reverse is also true: when you pass an electric current through a crystal of quartz it changes shape ever so slightly. However, the amount of electricity produced in this way is absolutely tiny, and anyway, quartz crystals are not squeezed during crystal therapy.

Even if the piezo-electric effect were having some unspecified upshot it could not explain the claims of crystal healing, as quartz is one of the very few materials with this property. Other minerals used for crystal healing, such as jade, amber or hematite, are not piezo-electric. No wonder geologists scoff.

○

Complementary medicines are often sold on the basis that they carry fewer risks than allopathic medicine. Its certainly hard to see how taking a homeopathic remedy can do any damage if there is nothing in it. Likewise, having shiatsu or reflexology sessions is not high on the list of hazardous pastimes. Acupuncture does carry some well-known risks, most obviously infection from poorly sterilized needles. This has been largely eradicated with the adoption of single-use sterile needles similar to disposable hypodermics. Needles have broken off inside patients and, among other things, punctured the spinal cord. Dr James K. Rotchford, writing in 1999 in the journal *Medical Acupuncture*, published by the American

Academy of Medical Acupuncture, reported that five deaths had been attributed to acupuncture. Nevertheless, when properly practised acupuncture carries a relatively low risk.

Other therapies also carry risks. For example, osteopathic or chiropractic manipulation can be extremely dangerous if performed on someone with a hairline fracture of the neck. The risk is low if the treatment is carried out by a properly qualified and registered practitioner, but it would not be accurate to call the treatments entirely safe.

Herbal medications can carry quite substantial risks, particularly if combined with prescription remedies. The popular herbal anti-depressant St John's Wort should not be mixed with prescription anti-depressants, as this can cause confusion, tiredness and weakness. St John's Wort also has the effect of lowering the efficacy of the contraceptive pill and so can increase the likelihood of an unplanned pregnancy. Mixing gingko biloba, sold as a memory enhancer, with blood thinning drugs can cause hemorrhage, as gingko also acts to thin the blood. Being 'natural' does not equal being safe, and that there are genuine risks from so-called natural herbal preparations and supplements.

It is probably true to say that, with the exception of some herbal remedies, the potential dangers of these therapies is less than that of many medical interventions. No complementary therapist is planning to perform surgery or has the range of potentially lethal drugs that an orthodox doctor has at their disposal. Proper training and regulation can minimize the risk and the increasing trend towards licensing of many different complementary therapies should contribute to safety.

There is, though, a way in which a therapy that has absolutely no physiological effect whatsoever can worsen health, or even kill.

This problem arises where there is a lack of communication between orthodox and complementary therapies. Many people turn to alternatives when they find allopathic medicine wanting. This is often for chronic conditions where conventional medicine has little to offer in terms of cure. Patients with diseases such as rheumatoid arthritis are prescribed drugs that they will have to take indefinitely and that, in the main, might reduce the symptoms of the disease but not actually arrest its progress. These could include painkillers or steroids, both drugs that can have powerful side-effects. The prospect of remembering to take such drugs every day for the rest of your life is not an attractive one and it is unsurprising that patients will look for alternatives. It may be that they find some complementary therapy that helps them feel better for whatever reason and tempts them to give up taking the drugs. This can be a major mistake.

Reputable complementary or alternative therapists do work alongside their patients' doctors and are very clear about their clients continuing to take their orthodox treatment, whatever it might be. However, having two arms of treatment working alongside each other produces a climate in which people move from one to the other in a way that could do them harm.

A good example of this hit the newspapers in the UK at the end of June 2004. Prince Charles, heir to the British throne and a famous advocate of complementary therapies, has lent his support to the Gerson Therapy. This is an anti-cancer regime that involves large quantities of fresh vegetable juice, a vegetarian diet, injections of liver extract and vitamin B12, and five coffee enemas a day. Cancer experts agree that diet is important in cancer and that plenty of fresh fruit and vegetables is a good thing. However, the Gerson regime also advocates giving up anti-cancer chemotherapy. These drugs can be very powerful and the side-effects deeply unpleasant, but they are used with care and expertise by cancer specialists and their track record in treating disease is well known. The Gerson Therapy

has no such medically established track record and many
oncologists have warned that it might be dangerous. The
American Cancer Society's web site states:

> There are a number of significant problems that may
> develop from the use of this therapy. Serious illness and
> death have occurred from some of the components of the
> treatment, such as the coffee enemas that remove potas-
> sium from the body leading to electrolyte imbalances....
> Some metabolic diets, used in combination with enemas,
> cause dehydration. Serious infections from poorly admin-
> istered liver extracts may result.... Relying on this treat-
> ment alone, and avoiding conventional medical care, may
> have serious health consequences.

The next thing that gets critics of complementary therapies
foaming at the mouth is the evidence for its efficacy – or rather,
the lack of it.

Evidence for the effectiveness of a medical treatment is
gathered in clinical trials. Broadly speaking a treatment is
compared with either another treatment or a dummy treatment of some sort. The quality of the data obtained depends
on numerous things, including the number of patients
involved in the trial, whether either the patients or the doctors taking part knew who was getting what treatment, how
the patients were selected, and how they were divided
between the trial group and the one receiving the established
or dummy procedure. It is a complicated and difficult process, of which more later. Essentially, a well-designed trial produces reliable data.

Normally, data from lots of different trials are assessed together in what is called a systematic review – a trial of trials. Systematic reviews allocate more weight to good quality trials than bad ones.

The largest and most respected library of systematic reviews is produced by the Cochrane Collaboration, an international network of experts in the analysis of clinical trial data. These reviews, known as Cochrane Reviews, are published quarterly and are available to all to read online. They are accepted as a reliable source of data about all sorts of different treatments and make a significant contribution to which drugs and interventions are used by doctors every day.

Two other sources of information about which treatments are effective for which conditions are the Internet journal *Bandolier* and a publication from the Department Complementary Medicine at the Peninsula Medical School, University of Exeter and Plymouth on the south coast of Britain, called *The Evidence So Far*. This booklet documents research conducted by Professor Ernst and his colleagues between 1993 and 2002 into a variety of complementary therapies.

It is an interesting exercise to search for data on studies into complementary medicine in these three sources.

Let's start with acupuncture, which has perhaps the best publication track record of any complementary therapy. Out of nine Cochrane Reviews for different types of acupuncture-based treatments, three are mildly positive, drawing conclusions such as:

Overall, the existing evidence supports the value of acupuncture for the treatment of idiopathic headaches. However, the quality and amount of evidence are not fully convincing.

or:

This review has demonstrated needle acupuncture to be of short term benefit with respect to pain, but this finding is based on the results of 2 small trials.... No benefit lasting more than 24 hours following treatment has been demonstrated.

The other six reviews all make comments such as:

The quality of the included trials was inadequate to allow any conclusion about the efficacy of acupuncture.

or:

There is no clear evidence that acupuncture, acupressure, laser therapy or electrostimulation are effective for smoking cessation.

Bandolier was started as a reference source for pain treatment, and as acupuncture is often used in this way the journal has many references on it. These are peppered with qualifying phrases such as:

Perhaps the biggest problem is that these trials, as a group, have avoided the hard question of longer-term outcomes. Even if acupuncture provides short-term relief, its place in management of back pain remains unknown.

or:

People entering trials of smoking cessation want to stop smoking. Some of them succeed. With acupuncture, no more succeed.

What of *The Evidence So Far*? It too considers acupuncture for lower back pain and concludes 'Acupuncture was shown to be superior to various control interventions, although there is insufficient evidence to state whether it is superior to placebo'. It also has a report on acupuncture to stop smoking and once again the results make poor reading for advocates. 'The results with different acupuncture techniques do not show any one particular method to be superior to control intervention'.

A similar exercise for homeopathy, spinal manipulation (chiropractic and osteopathy), reflexology and so on yields very similar results. Data tends to be low in quality and the conclusions equivocal at best. Compare this with a Cochrane Review of the use of aspirin for acute pain:

> Aspirin is an effective analgesic for acute pain of moderate to severe intensity with a clear dose-response.

The language is careful but the message is plain: aspirin stops pain. Compare this to the hedging of the reviews of acupuncture and it becomes apparent that it really does not conclusively cut it as a therapy.

Whatever advocates of complementary therapies might argue, the current evidence for their efficacy just does not stand up well to careful scrutiny.

Nonetheless, the proper application of science is already starting to bring the critics of alternative medicine closer to its advocates and to offer the prospect of health care that has the best of both approaches. The major stumbling block at the moment is the way in which evidence for a treatment's effectiveness is gathered and used, whether it is an orthodox or complementary approach.

5

the gold standard tarnished

Two crucial questions have to be asked about any medical intervention – will it improve the patient's condition and is it dangerous? If the answer to the first is yes and the second no, then it should be applied. If the answers are the other way around then clearly the treatment should not be used. Unfortunately, life, and health care, is never that straightforward. For virtually every treatment used in medical practice today the answer to these two questions tends to be 'probably' and 'not very'. Even the most common over-the-counter drugs, such as ibuprofen, do not remove every ache and pain and carry risks. Sometimes these risks are quite severe: ibuprofen can induce an attack in one in seven asthmatics and it is not possible to tell who is going to be affected.

Any intervention carries some risk and no treatment is guaranteed to work, so what patients and doctors need is a reliable measure of how effective a drug or procedure is to set against an assessment of how likely it is to cause a problem. This risk–benefit analysis is actually independent of treatment mechanism. Determining whether something is effective is not the same as understanding *how* it produces that effect. So in theory, any form of analysis that weighs up benefit against risk should be able to test any procedure, be it chemotherapy, massage or policy.

In medicine, the efficacy and safety of a treatment are assessed by clinical trial (or more normally by a series of clinical trials), and it is the data from these that determine whether a new treatment is to be accepted and used.

Giving a drug to one sick patient tells you very little about how useful a treatment it is. If the condition was getting better anyway, you could end up ascribing the improvement to the drug. If the disease was fatal, you could wrongly conclude that the drug was deadly. If nothing happened you might decide the drug was ineffective, even though without it the patient would have worsened – but you would never know.

In short, you need to be able to make comparisons with patients who did not receive the drug. In fact, a clinical trial is a collection of sophisticated compare and contrast exercises. The design, administration and analysis of a clinical trial are complex and there are a large number of researchers whose main interest is to examine and improve the way they are conducted. A badly designed clinical trial will reveal very little of value and is a waste of time and money.

Andrew Moore is the editor of *Bandolier*, the online journal of evidence-based medicine. His role is to evaluate clinical trials and assess the conclusions drawn from them in order to provide clinicians with the best information available on how to treat patients. He uses three broad criteria for assessing clinical trials – quality, size and validity. To design and execute a trial that conforms to these criteria can be demanding and not at all straightforward.

A control is crucial. It is a fundamental principle in all experimental research without which it is impossible to draw any meaningful conclusions. A simple experiment to test the effect of water on seed germination would, for example, have two boxes of seeds, one watered and one not. The dry box would be the control. A significant result would be if the watered seeds grew and the unwatered (control) seeds did not. The control was there to ensure that the seeds wouldn't have grown anyway, and that the watering was the key. The principle is exactly the same for clinical trials.

The most obvious form of controlled clinical trial compares patients who are given an experimental drug and those who are not. Straight away difficulties appear. If you ask someone to take part in a trial and then give them nothing, they know that they are part of the control group. Consequently they may not expect to get better, which in turn can influence the way they fare. They may show no improvement precisely *because* they do not expect to improve. The obvious thing is to give some patients a placebo – a mock treatment that looks or feels like the one being tested but which has no active components. Such 'placebo-controlled' experiments are commonplace.

Perhaps the first published placebo-controlled trial appeared in 1931 in the *American Review of Tuberculosis*. The study was an investigation of the drug sanocrysin for treating tuberculosis. Patients with TB were divided at random into two groups, decided by the toss of a coin, and one group was given the drug and the other distilled water. The researchers wrote: 'The patients themselves were not aware of any distinction in the treatment administered'. Therefore, when the drug turned out to have a positive effect on the patients who had received it, the study could be cited as proof of the medication's efficacy.

Placebo-controlled trials can provide good quality data, one of Andrew Moore's evaluation criteria. There are problems with them, though, both in principle and in practice.

Ethical qualms have always dogged placebo-controlled trials. But a paper published in the *New England Journal of Medicine* in 1994 reignited the debate. The authors asserted that conducting a placebo-controlled trial when an effective treatment already exists is unethical. This, they argued, was withholding treatment from a sick patient and ran counter to doctors' avowed responsibility to treat patients to the best of their ability. To support the argument they referred to the Helsinki Declaration of the World Medical Association – the international benchmark of medical ethics. One of the results of this debate was that in 2000 the World Medical Association issued an amendment to the Declaration of Helsinki – 'the Edinburgh amendment', agreed at a meeting in the Scottish capital. This strengthened the position on the inclusion of human patients in clinical trials so that 'every patient entered into a research project should be assured of the best proven prophylactic, diagnostic and therapeutic methods identified by that study'. The debate has not stopped there. For example, a paper published in 2002 in the *American Journal of Bioethics* argued that a placebo-controlled trial is justified if patients are not exposed to any serious risk and if they volunteer.

Even if circumstances allow the use of a placebo, the practical problems of finding a suitable one remains. It is very easy to come up with a placebo for a drug. An identical looking and tasting sugar pill usually does the trick. It is far more difficult to come up with a placebo for a treatment such as surgery or psychiatric counselling.

Placebo surgery does happen. A trial conducted at the Baylor College of Medicine in the United States in 2002 compared two common operations for arthritic knees with a sham surgical procedure. The two real procedures were keyhole surgery where a narrow tube was introduced into the knee joint and the surgeon inserted the instruments down that tube. The sham was simply making small incisions on the knee to mimic the those used in the real operations. The researchers found no difference between the placebo group and the other patients.

This surprising result suggested that some people might be undergoing unnecessary operations on their arthritic knees. 'We have shown that the entire driving force behind this billion dollar industry is the placebo effect' said study leader Dr Nelda Wray. 'The health care industry should rethink how to test whether surgical procedures, done purely for the relief of subjective symptoms, are more efficacious than a placebo'.

The knee operation was a relatively minor, minimally invasive procedure and the condition it was treating was not life-threatening. The patients recruited for the study were told that they might receive a sham operation and had to write on their medical charts that they understood this. This put off a significant number of potential patients: 44% of those originally identified for the trial refused to take part. This is a potential problem in any medical research, but 44% drop-out is very high.

Clearly the designers of this trial were confident that a placebo operation was ethical, as was the ethics committee that would have approved it. Every hospital or clinic where trials take place has an ethics committee that decides which studies may go ahead. It might have been a different story with something like heart bypass surgery.

Heart bypass surgery – replacing clogged arteries in a patient's chest with healthy blood vessels from their legs – can help prevent heart attacks. Hypothetically speaking, if there were an alternative form of surgery then one way to test its efficacy would be a placebo-controlled trial whereby some patients underwent the new procedure and some were cut open but then sewn up again straight away. Sham surgery would most likely constitute withholding effective treatment, and is thus extremely unlikely to be approved. A viable placebo, while being possible here, would not be ethical.

Taking this one step further, there are some surgical procedures for which it is impossible to imagine a placebo – from something as minor as remedial work for an ingrowing toenail to something life-saving, such as removing a malignant tumour. These important everyday medical procedures are simply not amenable to testing by placebo-controlled trial.

Clearly another form of control is sometimes required. The most common approach is to compare a new therapy with an established one. This sort of trial has two groups of patients with the same or similar condition; one group is treated with the established therapy and the other with the experimental one. Ethical concerns are dealt with in that every patient recruited onto the trial is given a treatment that is either known to work, or is expected to. And the problem of devising a suitable placebo falls away.

○

Think back for a moment to the 1931 TB drug trial. There was another crucial element to it: patients were divided at random into two groups, one of which received the drug and the other the placebo. This element of randomization is probably the most important feature of a clinical trial.

Randomization removes the vagaries of human individuality and allows general conclusions to be drawn. Randomization avoids the possibility that all the sickest patients end up in the control group and the less sick in the test group, an arrangement that would skew the results one way or another. It minimizes bias due to individual variation in responses to the drug. It smooths out differences in temperament, diet, lifestyle or the way in which the patients take the drugs. Professor Sir Iain Chalmers, former head of the UK Cochrane Centre, says that randomization is *the* crucial element. 'Randomization is the only thing special about the trials. When you are comparing like with like you have to randomize.... If you want to skip that you have to explain how you have ensured comparing like with like and this is the Achilles Heel of trials that do not randomize'.

Combining the two elements – randomization and a suitable control – results in the randomized controlled trial. This is the internationally acknowledged best means of testing a medical intervention of whatever sort, be it drug, surgery or physical therapy.

This is not the end of the story. A randomized controlled trial is still open to another form of bias that comes from the doctors and patients taking part in the study.

In 1784 King Louis XVI of France ordered an inquiry into the phenomenon of 'mesmerism' or 'animal magnetism'. This is the supposed ability to affect a 'universal magnetic fluid' that runs through the body and produce trances and healing as a result. The inquiry was headed by American Founding Father Benjamin Franklin, and included the noted chemist Antoine Lavoisier – famous for discovering oxygen. One of the tests was on a group of blindfolded subjects. They were either given the 'magnetism' or not and simultaneously told the truth about what was going on or lied to. The inquiry found that people only 'felt the effects of mesmerism' when they were told it was happening, regardless of what they really got. This was one of the first recorded blinded experiments – literally in this case – where the subjects did not know whether or not they were receiving treatment.

Blinding is an important part of a modern randomized clinical trial. A patient is told at the outset that they are going to receive an experimental treatment or a control treatment as ethics demands. But to ensure that their expectations do not influence how they respond, patients are not told which they are getting. This is called a single blind trial.

A second source of bias derives from the doctor administering the treatment. A doctor might react differently if he or she knew that the patient in front of them belonged to the control group, say. Therefore doctors are also often kept in the dark about which treatment they are administering. This is a double blind trial.

○

The data that emerge from a properly randomized, properly controlled, double blind clinical trial are likely to meet the first of Andrew Moore's criteria for good research. They are more likely to be good quality data.

The next criterion is quantity. A randomized controlled trial produces an average of patients' responses to the treatment. The more patients taking part in a trial, the more likely it is that the results will produce a meaningful average. It is not always possible to get a large number of suitable patients for a study –

the condition might be rare, or the treatment might only be designed for a small subset of people. Studies of physical interventions tend to be much smaller than those for pharmaceuticals. It is quite possible to recruit fifty or sixty thousand people for a drug trial, such as those involving the cholesterol-lowering statins. Performing a similar number of operations to assess different types of heart bypass surgery is, logistically, very unlikely.

The more patients in a trial the lower the potential for errors and the greater the likelihood of obtaining meaningful results, that is results with a high statistical significance. To work out the minimum number of patients needed for a meaningful trial can be fiendishly complex. Considerations include the rarity of the condition under test, how many patients may drop out or not take the drug, and how many false negative results the researchers are prepared to tolerate. Ethics comes into play too, as if the sample size is too small to produce meaningful results then patients could be put through a useless exercise.

The last of Andrew Moore's criteria is validity. This refers to the way in which the data have been analysed and the conclusions drawn. If, for example, a study shows that a new painkiller produces a 4% improvement in 3% of patients it is clearly far less efficacious than one producing a 50% improvement in 70% of patients. Even if the trial is immaculately designed, double blinded and including thousands of patients, it could still fail to be valid. There are different ways of analysing data, all open to different interpretations. And data analysis, like all the elements of a randomized controlled trial, is still an evolving discipline, and there are no hard and fast rules for doing it well.

The randomized double blind controlled trial has become the gold standard by which medical interventions are assessed, the most convincing analytical tool in clinical research. It is not, however, all powerful. There are limits to what such a trial can discover and what it can be used to investigate.

○

The first constraint is time. A typical trial might run over two or three years, within which time each patient might actually receive the treatment for between a week and six months. This may be more than enough to investigate the effects of a new drug for pneumonia or a new type of appendicitis surgery, but is too short a time to study an intervention for a chronic condition. Take, for example, smoking.

It is as good as proved that smoking causes all sorts of very serious ailments including heart disease and lung cancer. The most powerful solution is to give up, yet this is extremely difficult to test in a randomized controlled trial. The effects of smoking can take years to appear and it is totally unfeasible to run a trial that would last long enough to see whether giving up smoking really does improve health. Patients would have to either smoke or not smoke for half of their lives, and the lives of the researchers, before sufficient meaningful data could be gathered. The same problem affects drugs such as beta-blockers that people often take for the rest of their lives.

There are good ways of gathering long-term data. Sir Richard Doll famously linked smoking and lung cancer by looking back at smokers' medical records. This retrospective study examined the effects of smoking on health *after* they had appeared. A clinical trial, by and large, looks for effects *as* they appear.

Another limitation of the randomized controlled trial is that it cannot pick up rare events easily. This is simply a result of the numbers of people who need to be involved for something unusual to show up. If, for example, a drug causes a side-effect in 10% of the population, that side-effect will, in theory, be likely to turn up in a trial of just 10 patients. If the effect appears in just 1% of patients then the trial will need to include at least

100 people. When you get down to side-effects that occur in very small percentages of the population, only 1 in 50,000 say, then only a very large trial would pick this up. Such rare events are more often spotted in retrospective studies than in randomized controlled trials.

The next limitation is the fact that one randomized controlled trial is not enough. Each only asks a very specific question and that might not be the only question hanging over the condition or intervention under investigation. Take, for example, antibiotic treatment for urinary tract infections, also known as cystitis.

Cystitis can be very painful. It is found in people of all ages, particularly young children. By the age of five 1.7% of boys and 8.3% of girls have had a urinary tract infection. Antibiotics are the preferred treatment and are relatively straightforward to trial. Yet to build up a good picture of the clinical use of antibiotics in cystitis several different trials have been required. The Cochrane Collaboration currently reports that 13 valid clinical trials have looked at the optimal duration of treatment for elderly women, ranging from a single dose to two-week courses. Eleven have examined the use of the antibiotic methenamine hippurate as a way of preventing urinary tract infections. Three trials have studied long-term courses of two different antibiotics in children. Ten have examined whether short or long courses of treatment were most effective at clearing up the infection in children. In all, there have been at least 37 different trials just to explore the use of antibiotics in treating one type of infection.

Each clinical trial can be seen as one point in a join-the-dots picture. Not until all the dots have been joined up does the full picture become visible. It is not necessary to join every single one to get an idea of what the picture might be, but the more you do the greater the definition. A few good clinical trials will provide a useful outline of how a treatment works for a particular condition, but to get a detailed understanding many trials are needed.

There is an important consequence of this. Clinical trials examine one element of the treatment at a time; therefore they are very poor at looking at multi-factor interventions – more common in medicine than it might first appear. People with high cholesterol levels might be put on statins for life, but they will also be advised about diet and exercise at the same time. All are known to influence heart disease, but a clinical trial could only look at these elements in isolation. In this case it is relatively straightforward to design a trial that looks at, say, changes in diet while holding drug levels constant. But it can be problematic if a treatment is made up of a many different factors that cannot be separated.

The UK government funding body, the Medical Research Council, acknowledged this problem in a discussion document published in April 2000. Called 'A Framework for Development and Evaluation of RCTs for Complex Interventions to Improve Health', it contained a hypothetical scenario to illustrate the difficulties that 'bedevil well designed research':

> ... what is a physiotherapist's contribution to management of knee injury? The package of care to treat a knee injury may be quite straightforward and easily definable – and therefore reproducible: 'This series of exercise in this order with this frequency for this long, with the following changes at the following stages'. However, the physiotherapist may have, in addition to the exercises, a psychotherapy role in rebuilding the patient's confidence, a training role teaching their spouse how to help with care or rehabilitation, and potentially significant influence via advice on the future health behaviour of the patient. Each of these elements may be an important contribution to the effectiveness of a physiotherapy intervention. If we now

hypothetically consider evaluating a specialist stroke unit, the physiotherapist is one potentially complex contribution in a larger and more complex combination of diverse health professionals' expertise, medications, organizational arrangements and treatment protocols that constitute the intervention of that unit.

The document suggests that observational studies could be introduced as part of the way of measuring the trial outcomes. That is, rather than just measuring changes in clinical symptoms, such as the reduction in swelling of an arthritic joint, researchers should attempt to observe how a patient behaves as a result of the study. These ideas of broadening the purview of clinical trials are central to exploring complementary medicine.

The randomized controlled trial really struggles to cope with any form of psychotherapy. Whether it is given by a psychiatrist, a clinical psychologist or someone trained specifically as a psychotherapist is immaterial. The problem is the nature of the relationship between the patient and the therapist.

Firstly, it is impossible to conduct a double blind trial with any form of therapy that involves talking. One might construct a session wherein the patient is unaware of whether they are getting psychotherapy or just conversation, but it is impossible for the therapist not to know. At best any such treatment can only be single blinded.

Secondly, psychotherapeutic treatments tend to take a long time. Clients are typically in therapy for months or years, making them difficult and expensive to track.

Thirdly, the quality of the treatment is based on many things: the experience of the therapist, the relationship between the therapist and the patient, and the preparedness of the patient to cooperate. In other words there is no standard psychotherapeutic intervention in the way that there is a standard dose of a drug. Recruiting sufficient patients gets round some of these issues, just as recruiting thousands of people onto a pharmaceutical trial will average out individual responses. But what remains is that each psychotherapeutic intervention is specific to the therapist and patient. It is a tailored treatment. What works for one may have little effect on another.

As I write, 5,685 clinical trials are under way around the world according to the *meta*Register of Controlled Trials, an international database developed and maintained by the publishing consortium Current Science Group. These trials are good at determining the efficacy of a single intervention for a single condition. The data they produce are the average responses of a group of patients. A single trial answers a single question and many are needed to build up a full picture of how effective a particular treatment is. To get reliable data large numbers of people need to be studied, the more the better. The trial has to be properly controlled, whether by comparing the intervention with a suitable placebo or with an existing therapy. Trials have to be conducted ethically and no patients must be exposed to unnecessary risk.

Double blind randomized controlled trials are poor at assessing long-term interventions. They are poor at evaluating treatments that rely on the therapist–patient relationship. They are unable to pick up rare events or easily evaluate treatments for rare conditions. It is very difficult to design them for complex interventions and for individualized treatments. These limitations are known and accepted by the clinical researchers who use them and the data that come from them.

○

So far this discussion of randomized controlled trials (RCTs) has stayed within the realm of orthodox medicine. As the most powerful and most persuasive tool in the medical researcher's armoury the RCT is also the one most commonly applied to complementary therapies. This is all well and good for some treatments: herbal remedies, for example, can be trialed like conventional pharmaceuticals.

But the majority of complementary therapies fall into the category of treatments that randomized controlled trials are poor at assessing.

Homeopathic remedies are pills, tinctures or creams and so on, and could, one might imagine, be tested in the same way as pharmaceuticals. But that is not the only aspect of the therapy. Hahnemann, the inventor of modern homeopathy, insisted that the practitioner treat the patient, not the condition. As a result, two patients visiting a homeopath with similar problems may emerge with very different remedies. It is possible to design a trial to investigate a single remedy for a single condition, but homeopathy's use of different remedies for the same condition makes it a complex intervention, and therefore difficult to measure with a randomized controlled trial.

Similarly, acupuncturists will use different acupuncture points on different occasions while treating the same patients for the same condition. Shiatsu is also a complex intervention. The points that a practitioner stimulates vary depending on the state of the patient at the time of the consultation. In the same way, chiropractic, osteopathy, the Bowen Technique and reflexology all have elements that push them into the realm of complex interventions.

Intrinsic to many complementary therapies is advice on diet and lifestyle. Traditional Chinese Medicine, on which acupuncture, reflexology and shiatsu are based, is explicit in talking about how nutrition can affect health. Foods are divided into 'hot' and 'cold' – nothing to do with their temperature. 'Cold' foods include watermelons, pears and spinach and are prescribed to 'de-toxify' the body; 'warm' foods, among them ginger and garlic, are used to 'boost energy' and 'assist the body to heal itself'. Homeopaths will advise on diet as well, but almost all insist on abstinence from coffee and peppermint, as these can interfere with the remedies. It is possible to design trials that account for dietary factors as well but it complicates matters.

But where RCTs really struggle when it comes to complementary medicine is the practitioner–patient relationship. Alternative experts spend far longer on consultations than orthodox doctors. An initial session can easily be one and a half hours long, and follow-ups normally last between half an hour and an hour. In that time the practitioner will quiz patients about their lifestyles and their psychological state as well as their health problems.

The practitioner will then tailor what they are doing depending on the response of the patient. This could be a simple 'ouch' as a masseur presses a tender spot, but often or not the feedback is more subtle. Acupuncturists are trained to be sensitive to a large number of different 'pulses' throughout the body. Shiatsu practitioners report being aware of 'energy' moving within a client's body. Whatever the debates about the origins or even the existence of these phenomena, such therapists maintain that they can only detect and work with them if they pay close attention.

The consequence of this direct interaction is that the process of healing, according to many complementary practitioners, is one of becoming aware of a patient's state and responding to it. The therapeutic relationship, therefore, is crucial to the effectiveness of most complementary therapies. And randomized controlled trials are poor at assessing complementary treatments that involve such a relationship, just as they are poor at assessing orthodox psychotherapeutic ones.

These problems are not exclusive to complementary therapies. Doctors argue for the centrality of the doctor–patient relationship in good medical practice and are sensitive to changes that threaten it. A standard consultation by a family/ general practitioner has many of the aspects that make complementary therapy sessions hard to scrutinize with randomized controlled trials. GPs vary the treatment offered on the basis of how patients responded to the last option in much the same way that homeopaths or acupuncturists do. Doctors offer advice on eating and exercise.

The fundamental difference is that, by and large, an orthodox doctor is offering a series of treatment options that are considered as individual entities and assessed as such. Complementary therapy considers all of these elements together. Splitting them apart destroys the treatment.

○

The difficulties of designing randomized controlled trials to assess complex interventions and the therapeutic relationship are not the only ones that bedevil research into complementary therapies. There is the associated and knotty problem of 'outcomes'. That is, by what result, topic, event – outcome – is the impact of the intervention under investigation measured?

The selection of outcome measure is a key part of the criticism that advocates of complementary therapies level at randomized controlled trials. 'Well, no wonder you think this therapy doesn't work', they say, 'because you are deciding its success in terms relevant only to allopathic medicine'. Yet these cavils too go beyond complementary medicine.

In the wake of fierce debates through the 1970s about the medical control of childbirth, sociologists Ann Oakley and Hilary Graham pooled the results of their separate research projects on obstetricians' and mothers' different frames of reference of the experience. They found that obstetricians see reproduction as a medical topic, with pregnancy and birth as physiological processes, and the pregnant or in-labour woman as a patient, just one of a number on their case load. Women, however, see bearing a child as a natural process integrated into other aspects of life. This was brought into stark contrast when each group identified what a successful birth meant to them. For obstetricians, a healthy baby and well mother defined a successful pregnancy outcome. For a mother the definition was more complicated. Certainly a healthy baby was of prime importance, but a satisfactory experience of the birth itself and events thereafter were often included in her definition of success. An obstetrician might consider a birth successful that a mother deemed deeply dissatisfying if for instance if she felt uncared for and just another patient on the maternity production line. Conversely, an obstetrician might help deliver a sick baby and so record it as a failure, whereas the mother might have felt cared for and supported throughout the experience and see the birth as a success.

Clinical trials are designed to look for improvements in clinical symptoms. So a trial investigating the ability of turmeric to speed wound healing would measure how fast lesions closed up. The researchers would agree a definition of what 'healed' meant, how long you would normally expect a wound to heal and what was a suitable wound to be included in the trial. If the wounds treated with turmeric healed faster than the non-dosed wounds then it would be likely that something in the spice did indeed speed healing. More accurately, the trial would have shown a correlation between applying turmeric and faster wound healing according to the criteria used in that trial. That may sound pedantic, but it is crucial. A clinical trial tests only the criteria it sets out to look for: it is not a general exploration of what happens. Long before any patients are recruited, every researcher involved has to agree what they will look for and what they will consider a success.

When alternative medicines are put under the RCT spotlight the criteria for success have usually been traditional ones, such as reduction in inflammation in joints for osteoarthritis. Many complementary therapists argue that this ignores other benefits of their medicine.

Up to half the visits to acupuncturists in Britain are for the relief of osteoarthritis. Yet a systematic review of 13 trials of acupuncture in osteoarthritis, written by Professor Edzard Ernst and published in *Bandolier*, concluded that there was no evidence that it was more effective than either sham acupuncture, which pretends to pierce the skin, or placebo acupuncture, which does so at the wrong spot.

Several trials tracked the degree of pain that patients experienced. As pain is one of the main things that osteoarthritis patients complain about, it is an obvious measure of a treatment's success. Some researchers argue that it is not that simple.

○

John Hughes is a diffident, slightly earnest PhD student. We met at a conference on Developing Research Strategies for Complementary Medicine and got chatting over lunch as he waited nervously to deliver his short talk. He is doing qualitative studies of rheumatoid arthritis patients' perceptions of acupuncture. Rheumatoid and osteoarthritis have very different causes, but both give suffers life-long joint pain and both are difficult to treat with conventional medicine. Acupuncture is often given for both conditions, despite clinical trials finding 'no evidence' that is effective in either. Hughes believes that researchers might not be asking the right questions.

Qualitative research comes from social rather than natural science and has not always been as readily accepted in medicine as quantitative research. In a nutshell, qualitative research investigates what is happening, whereas quantitative research probes how much.

Hughes is doing a qualitative study of 22 patients with rheumatoid arthritis receiving acupuncture. He is conducting a series of open-ended interviews to find out what benefits patients *feel* they get from it. The people Hughes is questioning report an improvement in depression and/or mobility. Some admit they experience little or no pain relief, but say they feel more able to go about their daily lives. This hints that were clinical trials able to measure quality of life they might show acupuncture to be as useful as anecdotal evidence suggests.

The problem of outcomes in rheumatoid arthritis trials has been officially recognized by the OMERACT initiative. This is an international informal network of individuals, working groups and gatherings interested in rheumatology outcome measurement. The acronym originally stood for Outcome Measures in Rheumatoid Arthritis Clinical Trials, but they have subsequently dropped the arthritis clinical trials and changed rheumatoid to rheumatology. The outcome measures that the group issues are the *de facto* global standard for assessing clinical trials. The sixth OMERACT meeting in 2002 featured a major session on patient perspectives and recognized that these subjective experiences are not properly considered in rheumatoid arthritis trials. As a result, OMERACT is actively researching ways in which the outcome measurements themselves may be looked at in clinical trials of treatments for rheumatoid arthritis.

This type of research is relatively new in complementary therapy, and to date there are few completed studies. One such, published in January 2003 was commissioned by the European Shiatsu Foundation and entitled 'The Experience and Effects of Shiatsu: Findings from a Two Country Exploratory Study'. This was conducted by Andrew Long and Hannah Mackay of the University of Salford. It only looked at a small number of people: 15 patients and 16 practitioners in Germany and the UK.

The study had two aims: to 'uncover client and practitioner perceptions of the experience and effects of shiatsu' and to 'develop a protocol for the undertaking of cross-European cohort study of shiatsu clients'. Like Hughes' work on acupuncture it was attempting to lay the ground for a larger study by establishing how to measure the effectiveness of the therapy.

The researchers concluded that clients experienced a number of short-, medium- and long-term effects that they (the clients) felt to be beneficial. Long and Mackay reported the importance of clients having confidence in the practitioner and that the effects of shiatsu changed over time. They state that the study achieved its aims and, crucially, that it provides a 'base on which to design appropriate measuring tools for a wider study with larger numbers'.

Both of these studies are attempting to ask questions about what types of outcomes can be expected from two different types of complementary therapy. Neither is trying to find out whether the treatments are effective, just how better to define what they do. This is analogous to obstetricians asking pregnant mothers what they want from the birth of their child. The benefits of these types of study will not be limited to complementary medicine, but might also shed light on the experience of being treated by an orthodox doctor.

○

There is another, if not widely respected, form of clinical trial with which complementary medicine researchers are attempting to study individualized treatments and reactions: the *n* of 1 trial.

The symbol n in a clinical trial represents the number of patients involved; conventionally, the larger the number, the more reliable the results. One of the commonest criticisms of any trial is that the sample size was small – 'n was low' – and so few reliable conclusions can be drawn. Flying in the face of this logic, n of 1 trials are randomized clinical trials with just one patient.

A single patient with a condition is given a number, usually two, of different treatments or one treatment and a placebo over a period of time. Typically there are three pairs of treatment periods. The point of the trial is to determine which of the two treatments is better suited to that particular patient. It is not so different from a doctor trying out different treatments for a patient, just more formalized.

In a guest editorial in the December 2003 edition of the journal *Complementary Therapies in Medicine*, chartered statisticians Anna Hart and Christopher Sutton discuss n of 1 trials. For an n of 1 trial to work, they say, the treatment under investigation has to be fast to act and fast to stop working when ceased. If it is slow to act and slow to disappear its effects might spill into the trial period of the other treatment. Unfortunately, complementary and alternative medicines are often slow acting. This constraint can also rule out treatments that include lifestyle changes, such as major alterations in diet or exercise habits.

In the same edition of the journal a research paper combines the results of 24 n of 1 trials – on 24 different patients – studying valerian root and chronic insomnia. The researchers found no evidence that valerian does help insomniacs sleep, within the constraints of the trial. The important point here is not the result of the trial but the way it was conducted. As the editorial points out, combining 24 n of 1 trials is unusual and raises many methodological questions. However, in the way of science, the paper was published so that others could examine and criticize its methods and conclusions. It may be that n of 1 trials do not offer a solution to some or any of the problems of assessing complementary therapies, but until they are tried and scrutinized it is impossible to say.

○

There is little doubt that the effects of complementary and alternative medicines are often subtle – if not there would be little debate. As a consequence it is proving difficult for researchers to develop ways of testing them. I have outlined the intrinsic flaws of the randomized controlled trial – medical research's gold standard – and some of the ways in which researchers are trying to adapt it to the study of unorthodox medicines. This is difficult and is largely at the experimental stage. Debates still rage over the right questions to ask, the right statistics to use, the right trial designs and the correct outcomes. But that is the nature of research, if we had the answers there would be no need to investigate.

In November 2000 the UK Government House of Lords Select Committee published a wide-ranging report on complementary and alternative medicine. One of the appendices noted: 'Concerns over RCTs distorting a therapy or disguising its efficacy are not the unique concerns of CAM practitioners. Vincent & Furnham suggest that as attempts to apply the RCT to a wider and wider range of treatments have occurred, more and more problems have been uncovered. They list 10 such problems.... All these methodological issues apply to both conventional and CAM treatment trials. Therefore CAM is not necessarily a special case requiring radically new methodologies'.

There is an implicit coda to this statement. Research methods that provide answers for complementary therapies will have wider applications in medical research.

Much of the research into complementary and alternative medicines is actually research into how to do research. It is about developing the correct tools by adapting the ones that are available and inventing new ones. Another way of looking at it is that as medical research is testing alternative therapies, alternative therapies are testing medical research.

6

measuring the unmeasurable

Let's go back to John Hughes and his research on patients with rheumatoid arthritis. Hughes had noticed a discrepancy between clinical trials suggesting that acupuncture is ineffective and social survey data reporting that patients were happy with it. He began by interviewing acupuncturists to get an idea of what results they expected from treatment. Although he discovered a difference between traditional and Western acupuncturists, both were concerned to alleviate symptoms and to help patients to live with the disease. Hughes asked patients what they felt acupuncture had done for them. These two sets of interviews suggested to Hughes that the outcome measures typically used in clinical trials of acupuncture for rheumatoid arthritis are missing something.

Hughes describes the interviewing he uses as in-depth and semi-structured. He audio-tape records his interviews, transcribes them and analyzes them using grounded theory, a research approach introduced in 1967 by American sociologists Barney Glaser and Anselm Strauss. This is just one of the types of qualitative research beginning to be used to study complementary therapies.

Andrew Vickers is Assistant Attending Research Methodologist at the Memorial Sloan-Kettering Cancer Centre in New York. He has written extensively on complementary medicine research. In 1996 he published a commentary in *Complementary Therapies in Medicine* on a report which appeared the previous year from the UK's Nuffield Institute of Health, entitled 'Researching and evaluating complementary therapies: the state of the debate'. Vickers identified several 'myths' that the report had revived:

> It is claimed that 'more quantitative methods' are 'preferred by orthodox medicine' and 'more qualitative methods' are 'associated with complementary therapies'. No evidence is presented to support this claim. Not one

example of qualitative research in complementary medicine is quoted in the report. Though the case studies which are found in many complementary journals might be considered qualitative in nature, it is doubtful that any meet even a small number of the methodological criteria used to assess the rigour of such research.

The authors also complain about the low status of qualitative research in the hierarchy of evidence and recommend that 'research ... give greater credence to the use of qualitative methods.' It is not generally thought that qualitative research can, by itself, assess questions of effectiveness. Among other roles, qualitative methods can play an important part in determining the questions asked in quantitative trials and in helping to implement their results. However, they are not usually thought to produce direct evidence of effectiveness. The authors of the Nuffield report do not explain how qualitative methods could play such a role in complementary medicine.

In other words, qualitative research is not a cure-all.

All the same, there are calls from various quarters for qualitative work to take some sort of place in medical research, even if it cannot directly measure how effective a treatment is. The House of Lords' Science and Technology Select Committee on Complementary and Alternative Medicine is cautiously keen, although the report's discussion of the matter suggests that their Lordships, like many others, still have an uncertain grasp of quite what qualitative research actually *is*.

The discussion document from the Medical Research Council suggests taking a step-wise approach. They outline five phases of what they call a 'continuum of increasing evidence'. The main trial only happens at phase 4 and is preceded by work which includes teasing out elements of the intervention. Among the techniques available for this groundwork, the authors (a group of health service researchers and sociologists) list, a little inconsistently, various methods of qualitative data collection, including group interviews (focus groups), individual in-depth interviews, observational (ethnographic) research, preliminary surveys and organizational case studies.

'Qualitative research' the authors explain 'may save the researchers from making inappropriate assumptions as they proceed to design the next stage of the work.'

So even when tackling randomized controlled trials to evaluate complex interventions, qualitative research is recommended as a useful preparation.

A slightly different angle appears in a paper published in the *British Medical Journal* – 'Discrepancy between patients' assessment of outcome: qualitative study nested within a randomized controlled trial' – in 2003. This features an RCT of a suite of physiotherapy treatments and the provision of advice for patients suffering pain in the knees because of osteoarthritis.

The authors carried out in-depth interviews after treatment but before the main follow-up stage. They employed an experienced interviewer who used a checklist of topics to ensure that they covered the same ground with all 20 patients. Each interview was audio-tape recorded, transcribed in full and then analyzed independently by two researchers. To make sure these analyses were fair the researchers were blind to the answers that patients had provided to the quantitative questionnaires. Both quantitative and qualitative data included patients' assessment of their condition in terms of (1) improvement, (2) worsening or (3) no change in the pain that they felt or the extent to which their activities were restricted.

The authors describe their results as 'disquieting'. They found less than 50% agreement between the quantitative questionnaires and the qualitative interviews. They attribute this discrepancy to the circumstances in which the data were gathered. The quantitative data were collected in the presence of a doctor in the clinic where the trial was based. The qualitative data, by contrast, were collected by an interviewer who is not a health professional and in the patients' own homes – i.e. *their* 'territory'. This is a reminder that the context in which people say things affects what they say. The point, stress the authors, is that it is essential to get a good understanding of the variety of what the patient feels and thinks about a treatment when trying to find out whether an intervention is effective. And to do that the patients' viewpoints have to be studied in a way that is considerably more subtle than a quick questionnaire will allow.

Here qualitative research is *part of* a study. It is 'nested' into the trial rather than used as a stage in developing its design and is being put to work as a sort of a double check on the usefulness of quantitative data.

Calls for the inclusion of qualitative work are also, of course, echoed in alternative medicine research. In 2002 spiritual healer Su Mason and her colleagues Philip Tovey and Andrew Long, social researchers in health, published an article in the *British Medical Journal*'s 'Education and Debate' section. They point to some of the limitations of randomized controlled trials for complementary therapies and note that evaluations of complementary therapies need to include adequate attention to 'holism'; 'the intent to heal, non-judgemental listening... the healing environment, [and] users' expectations or attitudes' can and should be investigated by qualitative studies, they urge. They end their piece with a significant remark, almost a throw-away, but one that echoes a refrain that complementary therapists and their patients return to again and again: 'from the user's perspective, it is the beneficial effect itself that matters not how it was brought about'.

Complementary practitioners are not the only ones calling for qualitative research into complementary medicine. Qualitative studies, wrote Professor Bernadette Carter in the journal

Complementary Therapies in Nursing and Midwifery, can 'help researchers understand the meaning, beliefs and expectation that patients ascribe to CAM interventions'.

Randomized controlled trials, Carter notes involve a 'scientific, objective and detached' attitude to a study's outcomes. For complementary practitioners such an attitude is the opposite of their normal approach, which is to be 'subjective, engaged, helping, holistic'. On these grounds, she argues, it is inappropriate to introduce a study design which means that a practitioner has to go against their philosophy. Underlining her point, she also notes that patients have high expectations of complementary treatments. Her solution is to ensure that the research design reflects these features through qualitative research that explores the relationship between practitioner and patient. In her study of Bowen Therapy for frozen shoulder, she accommodated the therapists' desire for dependable evidence that would both be understood by allopathic practitioners and 'capture the "essence" of Bowen Therapy' for this group of clients. Complementary medicine needs, she says, to 'compete in the tough world that is allopathic medicine, they still have to be able to play the same games, only play them better'. It may be a challenge, but the line to pursue, she believes, is 'to fuse both qualitative and quantitative approaches'.

The reason that people have been calling for qualitative research is that it is relatively new as applied to medicine. However, a detailed 1998 review of the literature on 'Qualitative research methods in health technology assessment' by health care sociologist Elizabeth Murphy and her colleagues revealed that they are not a recent invention, dating back some 2,000 years. Hardly a new kid on the block!

○

Of the several types of qualitative research two in particular have been applied to studying complementary medicine. These stand in sharp contrast to one another. The first frequently uses in-depth, semi-structured interviewing along the lines of Hughes' study. Such interviewing is also sometimes called 'open-ended' and occasionally 'unstructured'. At times this approach may also involve observation. This divides into participant-observation and non-participant. In the former the researcher takes on a role relevant to the setting to be investigated, for example that of a clerical assistant or a hospital porter. Or it could mean just moving to live alongside the community under study. In non-participant observation the researcher avoids active engagement in any capacity other than observer. The second main type of qualitative data collection is conversation analysis, of which more shortly.

Qualitative interviewing is more like a conversation than a formal question and answer session with the interviewees encouraged to take the lead and interpret the topics as *they* understand them. Good qualitative interviewers do all they can to avoid imposing their own assumptions, offering instead maximum 'space' to the interviewees to express theirs. In particular, qualitative interviewing offers the interviewees the chance to raise issues that may not have occurred to the researchers. These discussions are usually audio-tape recorded (occasionally video-taped) and then transcribed for analysis.

These exchanges can be one on one. There can be two interviewers to one interviewee, with possibly a silent note-taking observer. Or they can be conducted in a focus group, a technique most often thought of as a market research tool but which is becoming more common in academic research.

Ursula Sharma, Director of the University of Derby's Centre for Social Research, wrote in her book *Complementary Medicine Today: Practitioner and Patients* about an unusual study she did of the use of complementary medicine in and around Stoke-on-Trent, an English town in the Midlands, away from the usual concentration on major metropolitan areas. She carried out in-depth interviews with people who used complementary medicine. Qualitative data, Sharma says, are 'probably more useful than large-scale survey data for this purpose, since they are better adapted to the study of process'. She acknowledges that repeat interviews with the same people over a long period could have been even better. She did not add that such studies are notoriously expensive.

What Sharma uncovered was not one simple decision to consult a complementary practitioner, but what she describes as 'chains of decisions taken over quite a long period'. The reasons her 30 interviewees first consulted echoed those of other studies: non-life threatening conditions, chronic conditions, and those where mainstream medicine had, in the eyes of the patient, been less than satisfactory. But the picture was more convoluted. People needed additional reasons to continue – or not – consulting a complementary practitioner, hence her evocative '*chains* of decisions'. She summarized these interlinked decisions by sketching three different types of user.

One she described as the 'earnest seeker'. This included those who had not yet found a cure, but intended to keep trying and those who were neither satisfied nor deterred. A young man declared, 'I have got very high standards for health'. He had used non-orthodox medicine for severe eczema for four years and had consulted spiritual healers, a reflexologist and an acupuncturist, and was having homeopathic treatment. 'Although he did not feel that his condition had improved very much as a result of all this effort, he appreciated certain features of the non-orthodox therapies he had used, especially the time given by the practitioners to diagnosis and their preparedness to discuss treatment with the patient', Sharma reported.

'Stable users' was her second type. These either regularly use some type of complementary therapy for a particular problem or use a single form of complementary medicine for most problems. One such had started to have homeopathy before the Second World War to deal with a severe ear infection. He and his wife had used homeopathy for most health problems ever since, even treating himself with it for routine ailments.

The third type Sharma labelled 'eclectic users' – those who used different forms of complementary medicine for all kinds of problems.

Sharma's use of qualitative research captured the multi-faceted nature of complementary medicine use, something that is far harder to pin down with fixed-choice questions typical of social surveys.

A small footnote to Sharma's work is worth reflecting on. Given that one of the features of complementary practice is held to be the quality of the relationship between patient and practitioner, and given that it is often characterized in terms of the length of consultation, it might be important to distinguish between the nature of the therapy and the context in which therapy is being offered. Complementary medicine is private medicine, i.e. a fee for a service, payable at the time. In Stoke-on-Trent, this was still cheaper than having private treatment via allopathic medicine. Research available at the time Sharma was working had already indicated that being able to 'buy' time was one reason for seeking private conventional medical care. One of her interviewees, educated and middle class certainly, but not at all wealthy, said:

The attention you get is important. Going private (for an earlier skin condition) I did get a chat and an explanation. I was treated as an intelligent person. It is worth paying for that, although you should not have to. Jenny (the herbalist whom she was currently consulting about her child's allergy) does explain what she is doing and I feel more active in the treatment.

○

Christine Barry, Senior Research Fellow in Medical Anthropology at Brunel University, argues that surveys rely on an idea of human behaviour as explicable with reference to internal, presumably psychological processes. Social anthropologists explain behaviour with reference to social interaction with others and pay attention to the context in which the behaviour occurred. A social survey conducted separately from the occasion in which the behaviour in question arose loses these features. In any case, Barry points out what is very well known: there is often a demonstrable discrepancy between what people say they did and what they actually did. Rather than use this discrepancy to discredit what people say, she instead found that inspection of this very discrepancy can itself be revealing. Actually watching what happens, preferably over a long time, as a supplement to interviewing, is the best way of exposing and exploiting this discrepancy analytically, Barry suggests.

Integrated Medicine is a concept that has emerged over the last decade or so. It attempts to integrate complementary and orthodox medicine and, hopefully, offer the best of both. Integrated, or integrative in the USA, medicine courses and centres have sprung up all across the developed world. The Association for Integrative Medicine claims members from Germany, Pakistan, Mexico, Brazil, England, Romania, Canada, Hong Kong and India. An example of qualitative research offers an interesting insight into integrated medicine at work.

Christine Barry completed her PhD thesis in social anthropology, a discipline which has qualitative research at its very core in 2003. Entitled 'The body, health, and healing in alternative and integrated medicine: an ethnography of homeopathy in South London', it is a wide-ranging study of homeopathy. One part focused on the work of a general practitioner who was trained in homeopathy among other alternative therapies. Barry observed and audio-tape recorded 20 separate consultations at a South London clinic. She also audio-tape recorded interviews with the GP, Dr Deakin (not his real name), and those patients she observed. The hours of tapes of consultations and lengthy semi-structured interviews were transcribed for analysis verbatim. Overall, she observed 23 consultations with Dr Deakin and 23 with non-medical homeopaths, conducted 46 interviews and filled 24 notebooks with fieldnotes: a huge amount of detailed data on actual practice in everyday contexts.

Barry found that Dr Deakin gave many of his patients the choice of treatments by asking them 'Would you like antibiotics, herbal or homeopathy?'. Some of his patients liked the fact that he offered alternatives to standard medical treatments, while others were confused by having to choose. There was no time in the brief consultations for him to explain the rationale behind each therapy, leaving some patients muddled about the homeopathic remedies they had been prescribed; many ending ended up not taking them. Barry contrasted these observations with her time spent studying non-medical homeopaths, who had a completely different philosophy. They conducted longer consultations, free from the constraints of a publicly provided health service, focused purely on homeopathic treatment. Their clients came to engage fully with the alternative philosophy of homeopathic medicine, she reports.

Barry concluded that Dr Deakin is a hybrid – neither fully homeopath nor general practitioner. He is a better listener and more empathic than many orthodox doctors, but, she felt, applies the principles of homeopathy as if they were just another type of treatment rather than a total philosophy of healing. He was also restricted by working within a standard general practice unsympathetic to alternative approaches. For example, he was refused the longer consultation times that are required for more holistic treatments.

Barry's work raises an important question. Does integration itself change the very nature of the type of medicine that a doctor practices? As she says, '...what version of a therapy is being integrated, how the patients approach it, what constraints are limiting its application and how does it differ in different contexts?'. One conclusion is 'for certain patients, choosing alternative therapies is in itself a reaction against orthodox medicine, and so the offering of alternative medicine by an orthodox system becomes a paradox'.

As with so much research, Barry's work raises more questions than it answers; there is, as yet, no answer to her paradox. However, it does show how qualitative research, this time from anthropology, can offer new insights into how any form of medicine is practised.

○

The other type of qualitative work is Conversation Analysis (CA). Conversation Analysts have for some years studied consultations in conventional medicine and are now beginning to turn their attention to complementary medicine sessions.

This is very different from the work considered so far. To provide a minimum illustration of what is involved, some technical detail is necessary.

CA was developed in sociology and linguistics. Paul Drew, sociologist at York University and one of the leading lights of the discipline notes, it is 'an observational science: it does not require (subjective) interpretations to be made of what people mean, but instead is based on directly observable properties of data... properties (which) can be shown to have organized, patterned and systematic consequences for how the interaction proceeds'. Conversation Analysis focuses primarily, but not exclusively, on verbal patterns which recur as people take turns talking with one another. It also takes into account features such as direction of gaze or body orientation. CA depends on *naturally occurring* interactions, those that happen whether or not any research is under way. Such interactions are captured by audio or perhaps video. The researcher does not, therefore, ever have to be present. Recordings are then transcribed in great detail. Everything said is transcribed verbatim, and other features, such as intonation, ums and ers, volume, and pauses are annotated. A series of symbols has been developed (see Table 1) allowing as faithful a transcript as possible of taped conversation to be typed onto the page.

There are three features to grasp to get a handle on CA. Any utterance and many non-verbal actions – body movements, say – are part of the conversation under investigation. These sounds or actions make up a sequence in which each is influenced by the one that came before. These sequences appear to have regular patterns.

Above all, a fundamental feature of conversation is that people *take turns* even though they do interrupt each other from time to time. This concept of *turn design* is central to CA and is readily visible in transcribed conversation. For example:

```
01   Dr:     Hi Missis Mo:ff[et,
02   Pt:                     [Good morning.
03   Dr:     Good mo:rning.
04   Dr:     How are you do:[ing
```

Table 1 Selected transcription symbols.

The relative timing of utterances

Notation	Meaning
0.7	Intervals within or between turns shown as time in seconds
(.)	A discernable pause too short to time
[ ]	Overlaps between utterances, the point of the start of the overlap marked with a single left-hand bracket
=	No discernible interval between turns. Also indicates very rapid move from one unit in a turn to the next

Characteristics of speech delivery

Various aspects of speech delivery are captured in these transcripts by punctuation symbols (which, therefore, are not used to mark conventional grammatical units) and other forms of notation, as follows:

Notation	Meaning
.	Falling intonation
,	Continuing intonation
?	Rising inflection (not necessarily a question)
:	Stretching a sound, the number indicate the length of stretching
Italics	Marks the sound stress
CAPITALS	Emphatic utterance usually with raised pitch
( )	Unclear or uncertain utterances speech placed in parentheses

```
05   Pt:                    [Fi:n]e,
06          (.)
07   Dr:   How are y[ou fe[eling.
08   Pt:            [Much [(better.)
09   Pt:   I feel good.
10          (.)
```

What is particularly nice about this example is the way that CA can illustrate how very similar phrases – *How are you doing* and

How are you feeling – uttered within seconds perform different actions. There is a slight difference in the construction of the turn (*doing* in the first, *feeling* in the second) and this reflects the contrasting actions which each performs. The first is an all-purpose general open-ended polite enquiry with which people frequently begin interactions. The second, though, is a medically focused enquiry inviting the patient to talk about whatever problem they have come to the doctor's surgery for. The patient distinguishes between the actions that each turn is designed to achieve. As Drew and colleagues observe: '(S)he responds to the former as a social enquiry with *Fine* (line 5) but to the latter biomedical enquiry with a form (*Much better. I feel good.* lines 8 and 9) which manifests her understanding that he is enquiring about her progress in coping with the condition about which she last consulted the doctor'. The point is that the patient's responses are connected to the designs of the preceding turns.

In contrast with other analytic approaches to audio-tape recorded conversations – be they interviews such as John Hughes did, or consultations such as Christine Barry analyzed – CA aims, say Drew's team, 'to identify and describe the specific interactional consequences which follow from given verbal practices'. CA typically entails very large collections of data: a study in the USA, for instance, included well over 300 consultations involving 19 physicians.

An example of the type of insight that CA can offer comes from the 2003 PhD work of John Chatwin, a student of Drew's, entitled 'Communication in Homoeopathic Therapeutic Encounters' and carried out at the University of York. Chatwin scrutinised 30 hours of homeopathic consultations recorded with 8 practitioners and 20 patients. He rarely found homeopaths directly telling a patient to stop taking an allopathic drug. 'It appears to be more common, when homoeopaths engage in talk about allopathic medicine, for categorical formulations to be attenuated and for comparatively subtle and sequentially extended approaches to be used', he comments. The homeo-paths, Chatwin's transcription indicates, set up hints so that it was the patients, by and large, who suggested giving up the drugs themselves. Chatwin's extract looks like this.

1 Hom: So how many cortizol ((allopathic drug)) have you got
2 left
3 (0.5)
4 Pat: Oh: not many=
5 Hom: =not many=
6 Pat: =no not many about (.) erm [(??)
7 Hom: [And how long are you going
8 to be doing this no- nasal spray for
9 (0.5)
10 Pat: Erm (.) I see the consultant (.) I mean I could stop
11 it now if=
12 Hom: =Yea
13 Pat: really shall I stop it now
14 Hom: Yea IF IT'D WORKED I'D'VE SAID <u>NO</u> Keep going with it but
15 it's NOT WORKED .h [then I wouldn't bother
16 Pat: [No I don't think it has. . .

From this Chatwin concluded 'the homoeopath would clearly prefer the patient to undertake a particular course of action – in this case to stop using her prescription drugs – she approaches the issue in such a way as to let the patient be the one to bring this into the open'. This does not appear to be evidence of deliberate manipulation; rather it reflects the patient-orien-tated approach of the homeopath. The meticulous approach of CA has highlighted some of the subtleties of the homeo-path–patient relationship. If such relationships are the key to alternative medicine's success, then maybe some hints to as to why lie in such studies.

There are signs that qualitative research into unorthodox medicine is being taken seriously. Christine Barry's post-doctoral work and three other qualitative projects are funded by the UK's Department of Health (following the House of Lords' Report) as part of a programme to strengthen research into complementary medicine. And debates about the standard of such studies are beginning to appear in the conventional medical literature. Helen Lambert and Christopher McKevitt, medical social anthropologists, writing in the *British Medical Journal*, warned of just picking up a qualitative research method and applying it without having a proper grasp of how to do so. They describe as misguided the habit of divorcing 'technique from the conceptual underpinnings'. In essence: the same rules apply to qualitative social science research as to medical research: understand the tools before you use them!

7

hints of understanding

It's time to stop dancing around the placebo effect and meet it head on. Clinical trials use placebo as a mark of failure; if something works no better than a sugar pill then it is not effective. Doctors embrace the placebo and cheerfully admit that it is an important part of their toolkit. Complementary and alternative practitioners say that their therapies work by encouraging the body to heal itself – a placebo effect – and critics argue that this means they do nothing. Nevertheless, research is beginning to suggest that the placebo effect may play a role in *every* act of treatment.

The word placebo comes from Latin, meaning 'I shall be pleasing'. The *Oxford English Dictionary* illustrates its use in medical terms thus: '...an epithet given to any medicine adapted more to please than benefit the patient'; 'It is probably a mere placebo, but there is every reason to please as well as cure our patients'. In other words, according to the *OED*, a placebo is not a useful treatment but a sop to a patient. These quotes, though, are from 1811 and 1888. Medical thinking has moved on a lot since then. In a paper published in the *American Journal of Pharmacy* in 1945 O.H. Perry Pepper described the placebo as 'the art of human care' and contrasted it with 'the science of medicine'. Complementary medicine researchers concur. Professor Paul Dieppe describes complementary therapies as being centred around the belief that 'caring for the patient' brings relief and is content that they may be largely based on a placebo effect.

Doctors have known about and exploited placebo for thousands of years. Hippocrates (c. 460–377 BC) clearly understood the power of the patient's mind. 'The patient, though conscious that his condition is perilous, may recover health simply through his contentment with the goodness of the physician', he said. Indeed, the Hippocratic Oath enshrines this understanding that it as important to care for the patients' general well-being as to treat them for their specific condition.

'... that warmth, sympathy, and understanding may outweigh the surgeon's knife or the chemist's drug'. Yet it is unethical, in the strictest sense, for a doctor to prescribe a sugar pill. And with good reason, as it is crucial that patients know what they are prescribed. Someone cannot give informed consent for treatment if they don't know what it is. Patients also need to know what drugs they are on in case they are taken ill and need to see an emergency doctor who doesn't have their notes to hand, or in case they consult another practitioner who needs to check for contra-indications.

In clinical trials placebo can have a huge economic impact: it is the standard by which newly discovered drugs are measured. One of the tests a drug has to pass on the way is whether it is more effective than a placebo in a clinical trial, or more normally in a series of trials that can include thousands of patients. In October 2000 the share price of UK-based biotech company Cantab Pharmaceuticals dropped 67% after its developmental drug to treat genital warts was found to be no better than placebo. A similar fate befell British Biotech when the first results from a programme of 10 randomized controlled trials cast doubt on its anti-cancer drug Marimastat. The *British Medical Journal* commented 'According to the industry, these huge falls illustrate how clinical trials are being closely tracked and the results used by financial analysts to make and destroy fortunes overnight. Analysts, they say, are now as likely to be found reading the *BMJ* and *The Lancet*, for clues about the progress of trials, as they are the *Financial Times*'. In financial terms, 'no better than placebo' can hurt.

It is inevitable, important even, that many drugs fall by the wayside during the development process. Companies know and accept this and normally have a portfolio of options under development to compensate. They are explicit about the risks, often including disclaimers alongside their development portfolio such as this one from the multinational giant GSK: 'Owing to the nature of the drug development process, it is not unusual for some compounds, especially those in early stages of investigation, to be terminated as they progress through development'.

For big pharmaceutical companies and their investors, placebo is firmly a negative. A drug that does not perform better than a placebo is a drug that will not make money. However, just because a drug does no better than a placebo does not mean that it does nothing. The simple act of administering a placebo does produce an effect.

Working out exactly what is going on is not simple, particularly when it is possible to produce a placebo effect without actually giving a placebo. A randomized clinical trial published in *Clinical Trials Meta-analysis* in 1994 found that if patients were told what to expect when given a drug they duly showed a bigger response to it. The act of giving the patient more information enhanced the effect of the drug. That is, the information alone elicited a placebo effect, yet no placebo was administered.

Some researchers refer to this placebo-effect-without-a-placebo as a 'context effect'. Context effects are the impacts of almost everything else other than the drug, surgery or whatever in question. They take into consideration all the factors within a consultation: the physical environment, the relationship between the patient and the practitioner, the amount of discussion between the two and the expectations of both. They can include everything from the moment a patient first becomes ill to the point at which they finish the treatment and stop seeing the doctor. Suddenly, what goes on between doctor and patient starts to look very complicated indeed.

Attempting to work out what is context effect and what is treatment is a significant challenge. A study that appeared in *The Lancet* in March 2001 identified 25 clinical trials that included an element of context effect. Author Zelda Di Blasi and her colleagues split these effects into two categories, cognitive care and emotional care. The former aims to improve the patient's understanding and expectations of the treatment; the latter seeks to make the consultation a more relaxed and less fearful experience. The researchers found the trials to vary considerably and did not draw any firm conclusions. The one consistent observation they made was that 'physicians who adopt a warm, friendly and reassuring manner are more effective than those who keep consultations formal and do not offer reassurance'.

What is treatment and what is context becomes even more blurred for complementary therapies. A common theme from homeopathy to shiatsu is providing a good environment for the patient, one that is welcoming, encouraging and unhurried. A typical therapy room has soft lighting, comfortable furnishings and perhaps a candle burning and soothing music. Therapists take time to make their rooms as pleasant to be in as possible. Complementary practitioners tend not to split environment from treatment. They see the place where someone receives their therapy as part of that therapy. In other words, practitioners deliberately seek out context effects. Contrast this with the orthodox medical view that a drug will be efficacious whether it is administered in a plush suite or on a battlefield. That is not to say allopathic doctors do not care about where they practice, just that environment is not as deeply linked to treatment in the same way.

○

A good argument for seeking a positive placebo effect is that it also has a dark side. Derived from the Latin meaning 'to harm', nocebo is placebo's evil twin.

This idea has probably been around for as long as we have. Concepts such as being 'scared to death' or 'putting a curse on' an enemy are well established in all cultures. Haitians believe a voodoo curse can kill – victims are not poisoned or injured, just told they are going to die. Sicilians believe that a hex can cause headaches and versions of 'evil eye' superstitions are found right across the Mediterranean, Middle East and Asia. There is far less research into nocebo than placebo, but what has been done is arresting.

An oft reported study was conducted on 57 Japanese school-boys to determine their response to allergens such as pollen and nuts. The boys filled out questionnaires relating to their past experiences with rashes and skin problems. The research-ers then selected those boys that had reported sensitivity to the lacquer tree, which, like poison ivy, gives some people a rash. The scientists then blindfolded these boys and told them that one arm was about to be brushed with the leaves of the lac-quer tree and the other with chestnut tree leaves. The research-ers then brushed chestnut leaves on the arm the boys thought would receive lacquer tree leaves and vice versa. Almost imme-diately, most of the arms that had expected lacquer tree foliage developed red, irritated rashes. The other arms remained unblemished – despite having touched the aggravating plant.

Another famous case was part of the Framingham Heart Study. This long-term study of cardiovascular disease was set up in 1948 and funded by the US National Heart Institute, now the National Heart, Lung and Blood Institute. Over 5000 inhabitants of the small town of Framingham in Massachusetts were recruited in the first wave; 23 years later another 5000 signed up. As the *Journal of the American Medical Association* reported in 1996, the project revealed that women who believed they were prone to heart disease were four times as likely to die from cardiac problems as women who had exactly the same risk factors but did not believe they were in any additional danger. The conclusion the researchers reached was that the mindset of the first group contributed to their ill health. They thought themselves ill.

Designing experiments to study nocebo is an ethical morass – to do so would be to set out to cause harm. In the case of the Japanese schoolboys the harm was small and so considered justifiable. The nocebo effect from the Framingham Heart Study emerged from data analysed retrospectively; clearly it would have been impossible to deliberately try to induce a life-threatening condition by suggestion.

Then there is the additional ethical question of informed consent. Part of informed consent is telling the participants of any known side-effects of the what they may or may not be getting. The implication from what we know about nocebo is that if you tell someone that they might experience tiredness and nausea, say, then they are more likely to report feeling tired and nauseous. This is particularly problematic when a new treatment is being compared with a well-established one. What should each group be told? Is it ethical to tell someone that they might get a certain side-effect when it has not been reported for that drug but instead for the experimental medication? Does that then produce an artificial set of side-effects that continue to be ascribed to the new drug? No one knows.

○

Much of the evidence for placebo, nocebo and context effects comes from clinical trials, which, by their very nature, do not offer explanations for how the effects appear. Implied in all of them is the fact that thoughts or state of mind can have an effect on health, one of the tenets of all complementary medicines. Much of modern medical research has its origins in the mechanistic model of Cartesian duality – splitting the mind from the body and treating it as a machine. The truth is not that clear-cut. Doctors certainly recognize the importance of a patient's state of mind and the literature is full of papers reporting it and discussing the implications. A major contribution comes from a new branch of science that is finding evidence of direct links between mind and body.

By the end of 1970s there were enough immunologists, psychologists and neurologists with similar interests working in similar fields for the discipline of psychoneuroimmunology to be born. The phrase was coined by Robert Ader, who was responsible for some of the seminal work in the field now best described as the study of how state of mind influences the immune system and vice versa. Psychoneuroimmunology is now an established discipline with its own journal and a professional society that has met annually since 1993.

Broadly, our immune system has two roles: to fight off infection and to identify and destroy rogue cells before they can turn into cancers. It is often likened to a defending army charged with both repelling boarders and putting down insurrections. Like an army it has different weapons that can be applied in different ways. Similarly, it has an elaborate control system designed to keep it in check, because an uncontrolled immune system is just as dangerous as an uncontrolled army. Fast-moving mobile scouts spot the enemy; slower but more powerful heavy artillery can be called to the trouble spot.

When the immune system's scouts detect an invading bacterium or virus, they send out signals summoning the heavy squad to kill the invaders. These signals are chemicals called cytokines. Amongst other things, cytokines increase the flow of blood to a damaged or infected area and make blood vessel walls leaky so that white blood cells, the heavy artillery, can get to the trouble spot. Having made a problem area warm, red, swollen and painful, cytokines attract reinforcement white blood cells as long as the infection or damage persists. The cytokines involved in the inflammation response – pro-inflammatory cytokines – are central to current research in psychoneuroimmunology.

Over-production of one pro-inflammatory cytokine, Interleukin-6, is associated with many chronic degenerative illnesses, including cardiovascular disease, osteoporosis, arthritis, type 2 diabetes, cancers, gum disease and general frailty and decline. A study published in 2002 in the *Proceedings of the National Academy of Sciences* demonstrated how chronic stress effects levels of Interleukin-6. Janice Kiecolt-Glaser and co-workers identified 119 people who were caring for partners with dementia and compared them to 106 similar people without caring responsibilities. These were sufficient numbers for a valid conclusion. The carers produced on average four times as much Interleukin-6 as the non-carers. Surprisingly, this level of overproduction continued for some years after their spouses died. The team argued that these data offered a mechanism by which living with stress could translate directly into chronic diseases.

This study is part of a larger body of research that connects social situation with well-being. One rather elegant paper showed a strong link between having a good network of close friends and relations and the ability to fight off colds. Another found a correlation between high levels of stress hormones in the first year of marriage and the likelihood of divorce. All told, psychoneuroimmunological research has demonstrated that exercise, stress, depression, sleep, social isolation, accident or bereavement can have effects on health issues from susceptibility to infection to cancer. And the traffic between mind and body is two-way. Glitches within the immune system can feed back to the brain and precipitate changes in mood and perception.

Alongside psychoneuroimmunology are other studies that are beginning to reveal ways in which specific placebo effects might work. Some of the most compelling have been carried out on patients with Parkinson's Disease, the degenerative neurological disorder that affects walking, writing, speech and so on. Typically there are three main symptoms: tremors, muscle stiffness and slowness of movement. People with Parkinson's can also get depressed and tired, and find it difficult to communicate. It is a cruel condition that slowly and relentlessly gets worse.

The disease is caused by the gradual death of the cells in the brain that produce the essential chemical messenger dopamine. There is no cure, but taking synthetic forms of dopamine can help and there has been some success with transplants of dopamine-producing cells into the brain.

Doctors working with Parkinson's patients have noticed that in some stressful circumstances, such as the need to escape from a fire, normally immobile patients can become briefly mobile. Two suggestions have been put forward for how this 'kinesia paradoxica' might work. The body might have an emergency escape system that does not rely on dopamine, or stress may cause a burst of dopamine in the brain. Thought, that is to say, can have a direct metabolic effect.

Alongside this phenomenon there have been a number of studies that indicate that the placebo effect is particularly strong in Parkinson's patients. It is often the case, for example, that those in the placebo groups of clinical trials show a significant improvement.

A series of experiments from the Pacific Parkinson's Research Center at the University of British Columbia has suggested a mechanism for these observations. They have also provided hard evidence that administering a placebo can cause physiological changes. The researchers used a type of brain scanning called Positron Emission Tomography, or PET. This technique measures the amounts and locations of chemicals in the brain – in this case, chemicals associated with dopamine release. One group of subjects was given a common Parkinson's drug and the other an injection of saline. Both groups produced a surge of dopamine in response. Those given the drug had a slightly bigger surge, but the differences were relatively small.

The researchers propose that this placebo effect works through the brain's reward mechanism. This general brain circuit unleashes dopamine in response to or anticipation of activities such as eating, having sex, or some other behaviour producing a feeling of well-being. In the main it is a valuable encouragement to animals and people to do things that are good for them; if there was no reward for eating or copulating then the species would rapidly die out. The circuit can be hijacked though. Addictive drugs like alcohol, heroin and cocaine stimulate dopamine release.

The reward mechanism appears to be tickled by the anticipation of receiving the treatments, not receiving them. The Parkinson's patient anticipates that the placebo injection will have an effect, and this activates the reward mechanism which produces a surge of dopamine. It could also explain kinesia paradoxica. The anticipation of getting away from a source of stress may activate the production of the dopamine that the brain needs to tell the muscles to work.

This neat piece of research offers a clear biochemical explanation for one placebo effect. But as one of the researchers pointed out in the journal *Science* in August 2001, these results 'suggest that in some patients, most of the benefit obtained from an active drug might derive from a placebo effect'.

The dopamine reward pathway is not necessarily a general mechanism for all placebo effects and it is probable that different conditions may have a different susceptibility to placebo. There is almost certainly a variation in the way different people respond to placebo as well. Nonetheless this research demonstrates two things: that the placebo effect can be real, measurable and physiological and that it can make a significant contribution to the improvements that patients show after treatment. Likewise psychoneuroimmunology does not provide a universal explanation for why treating a patient well makes them better. It merely provides evidence of a link between social condition and health. Both these areas suggest some important questions for future studies to tackle. Not least, do complementary therapies elicit physiological effects similar to those seen in the experiments above?

○

George Lewith is Senior Research Fellow at the University of Southampton and Visiting Professor at the University of Westminster. He is a medical doctor and an experienced acupuncturist, and has been researching alternative medicines for years. He describes one of the key elements of a complementary consultation as 'listening with an intent to heal'. This, he argues, is fundamentally different from just providing a sympathetic ear.

Complementary consultations range from 30 to 90 minutes, compared to the seven to 20 minutes or so that a general practitioner offers. Lewith, though, contests the idea that if every doctor could spend an hour with each patient then they too might make better headway in seemingly intractable situations. This, he argues, is to misunderstand the nature of the therapist's role. If a patient has a chronic condition that orthodox medicine cannot treat then a conventionally trained doctor might listen for three-quarters of an hour feeling impotent in the knowledge that there is nothing by way of treatment to offer. By contrast, says Lewith, an acupuncturist might not feel they can cure the patient, but may well believe they can make them feel better. So they listen with the hope and expectation that they can do something. While the patient tells their story, they will be thinking 'I wonder if spleen 6 would work here' or 'There has been an invasion of damp, what can I do about it?'. It is an active, participatory process, different, Lewith contends, from a passive orthodox doctor spending time just for the sake of it.

It is a hard idea to test, but Lewith is running a study to see if he and his co-workers can identify an effect from 'listening with intent'. The team is recruiting women with undiagnosed severe pelvic pain. This chronic condition can ruin lives. These women have seen specialist after specialist yet are still suffering. All attend a pain clinic and have had considerable amounts of specialist medical intervention. This, says Lewith, suggests that it is not just a matter of time with a doctor that counts. After recruitment these women will be given a Traditional Chinese Medicine consultation and diagnosis. Crucially, the women will be offered an explanation of their condition in Traditional Chinese Medical terms. This diagnosis is unlikely to have any grounding in modern medical science, but the point is to investigate whether providing a model for understanding can influence the condition.

Lewith also plans a secondary study to ask further questions about what, as he describes it, 'sets people up for success'. Is it belief, depression or state of mind? Is it possible to separate the process of diagnosis from that of treatment? He is trying to find out what, if anything, is special about the way in which a complementary practitioner practises compared to an orthodox doctor.

○

Our growing understanding of placebo, nocebo and context effects is challenging to some complementary practitioners. It

implies that what they achieve has nothing to with the treatment they offer and everything to do with the way they do it.
More positively, it hints that context effects need to be carefully monitored, nurtured and controlled to reach their maximum efficacy. Throwing someone into a cold, stark room and
sticking needles into them without any preamble is far more
likely to elicit nocebo than any benefit. Medical research may
never find any other effect than placebo for some complementary and alternative medicines. But if the placebos they produce are bigger, more powerful and more effective than any
others then these too could be considered powerful healing
methods that medicine might wish to harness reliably, rather
than dismiss.

The nocebo and placebo phenomena remind us of the
complexity of healing and the importance of the art of
medicine.

Malcolm P. Rogers MD, 2003

8

a new model

The rise of testing of complementary therapies has mirrored but lagged behind the growth in their popularity over the past two decades. Clinical trials of various types have probed orthodox treatments over the past 50 years or so but there is a far shorter tradition of using them in alternative medicine. Then there are all the added subtleties of complementary therapies: outcome, context, therapist–patient relationship and so on. The result is a relatively small group of people familiar enough with all the threads to draw them together coherently.

One such is Andrew Vickers of New York's Memorial Sloan-Kettering Cancer Centre. In 1996 he published a paper in the *Journal of Alternative and Complementary Medicine* entitled 'Methodological Issues in Complementary and Alternative Medicine Research: a personal reflection on 10 years of debate in the UK'.

Vickers outlined his view on the shortcomings of conventional medical research techniques in the study of alternative therapies. He did not call for a whole new set of tools, instead arguing that randomized controlled trials can be adapted. He summarized his thoughts in a table that matches research questions with study design.

The question that a researcher wants to answer is the key to all good scientific endeavour, whether in anthropology, physics, genetics or acupuncture. A clear, well-formed query will lead more easily to useful results. Much of the challenge of any project is formulating that initial question. (Incidentally, good qualitative data on complementary medicine should help define good questions for quantitative research.)

Vickers took acupuncture as a model therapy and tabulated 10 types of research question that might be asked about it and the type of study best suited to answering them. For example:

Question: Is acupuncture an effective treatment in practice?
Research Design: Audits of practice with long-term follow-up using validated outcome measures; comparative cohort studies; pragmatic research: randomized trials comparing acupuncture as a package of treatment to standard care.

Question: Is acupuncture a placebo?
Research Design: Fastidious, randomized, placebo-controlled, blinded trials.

The rest of the table covers other questions such as what type of conditions acupuncturists treat or what patients experience. Each draws on a different set of research methods. It's a lucid illustration of how different methods can offer different answers and why using more than one method is essential to getting a complete picture.

One of the leading proponents of properly controlled clinical trials of complementary medicine is Edzard Ernst of the Universities of Exeter and Plymouth. He trained and practised as a doctor in Germany and studied a raft of complementary therapies including acupuncture, autogenic training, herbalism, homoeopathy, massage therapy and spinal manipulation For the past 10 years, Ernst has been conducting research into all forms of complementary medicine and has published prolifically in the field. He and his colleagues have compiled the booklet *The Evidence So Far*, mentioned previously; it is a synopsis of clinical trials and studies that the Department of Complementary Medicine has either done or reviewed. A broad document, *The Evidence So Far* covers many different areas, from osteopathy, chiropractic and acupuncture to homeopathy.

The key questions in complementary medicine according to Ernst are those of medicine as a whole, namely: 'Is it safe?' and 'Does it help anybody?'. Ernst is evangelical about the power of rigorous research to answer these questions and is adamant that good research comes from collaboration. It is no good attempting a trial in acupuncture, he says, unless there is an acupuncturist, an expert in trial design and a statistician on board from the outset – it is rare to find all these skills within one individual. However, to design such a trial is expensive; it costs money just to convene the investigators before any patient recruitment even starts. Ernst reckons that one reason why so many complementary studies are weak is lack of money. A major consideration for the future of this type of research is funding. Some money comes from the manufacturers of herbal medicines, aromatherapy oils and flower remedies, but it is unreliable and paltry next to the huge sums spent by big pharmaceuticals companies.

His question 'Is it safe?' is only partly answered by randomized controlled trials. They pick up obvious hazards, but the numbers of participants are too small to detect rare side-effects. It is better to do retrospective epidemiology by examining patient records. Again, Ernst emphasizes the need for rigour and teamwork.

Ernst has big concerns about the placebo effect. As a clinician he wants his patients to get better and concedes that *how* that happens takes second place. But as a researcher he finds it very difficult to sanction a treatment that appears to have no intrinsic effect. It is regressive – a huge step backwards into the dark ages of medicine – because it doesn't lead anywhere.

That said, Ernst does not disregard the placebo effect; he believes it should be studied in greater detail. Why a spiritual healer elicits a huge placebo effect where a doctor does not is a fascinating research conundrum. The phenomenon is important to understand, he says, as all practitioners want to exploit it.

He has mixed views on qualitative research. The good stuff, he recognizes, can answer important questions about the meaning of illness or suffering. But it has little to say about efficacy. Worse, he cautions, there has been a proliferation of poor qualitative research, often questionnaire-based. This can be dressed up as proving something that it is incapable of doing, he warns. He calls it 'politically correct' research that does not provide hard answers but does not upset anybody.

Professor Ernst's views have earned him opprobrium from many quarters, and he has the letters to prove it. Nevertheless he stands by his insistence that scrupulously controlled, properly designed clinical trials are powerful tools in the study of complementary medicine. This scientific approach has won over a number of medical consultants, he argues, who now approach the subject in a much more positive light.

Medical researchers and complementary therapists don't talk to each other enough, agrees Professor Stephen P. Myers, Director of the Australian Centre for Complementary Medicine Education and Research, a joint venture of the University of Queensland and Southern Cross University. Myers is curious about the influence of consultation and individualization on the effectiveness of homeopathy. Clinical researchers often design and conduct experiments into homeopathy, he points out, without ensuring that they are actually assessing what homeopaths really do. In an attempt to overcome this problem, Myers brought together trial experts and homeopaths for a brainstorming seminar. The study he is now overseeing is a direct result of this collaboration and is being undertaken by Don Baker one of his PhD students.

They have put together a set of osteoarthritis patients, confirmed by detailed diagnosis including X-rays of their deteriorating joints, who are prepared to take part in a placebo-controlled trial. To date they have recruited 115 of the 135 patients they seek to enrol. The patients are divided randomly into two groups; one has a consultation from a homeopath, one does not. These groups are then further subdivided. One arm of the no-homeopath group receives a complex of homeopathic remedies designed to treat osteoarthritis; the other arm receives a placebo. The consultation group is divided into three arms. The first receives the same homeopathic complex, the second a placebo and the third group an individualized homeopathic remedy.

This five-armed trial is designed to answer a number of questions. First: what is the effect of consultation on the efficacy of the homeopathic complex for osteoarthritis? This will show up in any differences between the groups that receive the complex and placebo, the only difference being that one set will have had a consultation and one will not. Second: what is the effect of individualizing homeopathic treatment? Any differences between the fifth group and the others will suggest that customizing impacts efficacy.

This is an elaborate exercise, designed by experts in clinical trials and homeopaths from the outset; it is an *experimental* experiment. It might not be perfect, but it should point the way to a reasonable next experiment. It is good science, in other words.

The idea that well-designed clinical trials can determine the efficacy of complementary medicine is echoed by Richard Nahin, Senior Advisor for Scientific Coordination and Outreach at the National Centre for Complementary and Alternative Medicine, part of the US Government-funded National Institutes for Health. He says 'the same techniques and methods used to study conventional biomedicine are applicable to complementary medicine'. Although he does point out that 'there may be some challenges in studying complementary medicine, such as randomization, blinding, and quality control, these challenges are not unique to complementary medicine'.

Nahin also recognizes the need for outcome measures to be relevant to what the trial is trying to do. In his view, 'If the purpose of the trial is to establish clinical efficacy, then state of the art measures of clinical efficacy specific to the disease or condition must be used. If the purpose is to assess changes in quality of life, accepted measures of quality of life must be used'. His position on the placebo effect, however, appears to be somewhat different from that of Ernst. Speaking on behalf the of NIH he said the organization as a whole is interested in 'interventions that can affect patient expectations and clinical outcomes regardless of their mechanisms of action.... In particular, we are interested in therapies that are not directly biologically active but that, nevertheless, produce changes in health by biological means through the mind–brain–body connection'.

Another approach is typified by the work of Aslak Steinsbekk of the Department of Public Health and General Practice at the Norwegian University of Science and Technology (NTNU) in Trondheim, Norway. He is a homeopath with a degree in sociology and so, perhaps inevitably, has a somewhat different viewpoint. In January 2004 Steinsbekk put forward a suggestion to the World Health Organization expert group on homeopathy. This described a strategy for research development for complementary therapies and other therapies widely in use but with a lack of research. This strategy is used by the National Research Centre on Complementary and Alternative Medicine, University of Tromsø, Norway, and can be compared to the development of pharmaceuticals that are to become licensed and even reimbursed.

The first stage in pharmaceutical development is the identification of a chemical that appears to have some sort of biological activity. Further experiments explore its mechanism, confirm its potential and lead to tests in animals. If these prove favorable the next stage is preliminary safety tests in humans. After this comes a series of steadily larger clinical trials to prove efficacy and effectiveness and to work out exactly how the drug should be used. Only then can it be licensed and released.

The path of a treatment that is in use on patients by the time researchers turn their attention to it, like complementary therapy, runs in reverse. The initial set of studies is to first describe the use; then they move on to safety before they determine the effectiveness of the therapy as it is given in everyday practice. If this turns out to be beneficial for the patients, then the efficacy of the specific components is investigated before finally moving on to the mechanism of action. The starting point for research into the effectiveness of pharmaceutical drugs is that they are an explanation looking for an application; complementary therapies are an application looking for an explanation.

When putting together a clinical trial, Steinsbekk's begins with the patient's perspective. Too many studies are done, he argues, to answer the questions that a researcher has about a therapy rather than what patients want from it. Results from studies so designed have no bearing on how patients use a therapy in real life. A trial might show that a particular homeopathic remedy has no effect on a heart condition, but if a patient wouldn't seek homeopathy for that complaint then so what? It's like spending time discovering that cotton wool makes a poor road surface – scientifically valid, but practically useless.

Also important, says Steinsbekk, is how patients are recruited into trials. For studies of orthodox medicine, subjects are often selected from specialist clinics. This doesn't work for complementary therapy, he argues.

Patients in a specialist clinic have usually been referred by other doctors. The first line of treatment probably hasn't worked for them and their condition is a little more tricky. This is very different from the group of patients that a complementary therapist might have on their books. These are likely a far more diverse group with an equally diverse set of symptoms and conditions. As a result, complementary therapists have to develop expertise in dealing with a wider set of problems. Clinic doctors and nurses, by contrast, are specialists.

For a study looking at the effect of a complementary therapy on diabetes, say, it would be only natural to recruit patients in the same way as for a drug study – from a clinic. The complementary therapists in such a trial would face a set of people with intractable diabetes rather than their normal cross-section of patients. They will be able to offer something from their repertoire, but it is not their focus. It is like asking a domestic central heating engineer to repair an industrial boiler. The engineer would make a good attempt at it, but probably wouldn't do as well as an industrial boiler expert.

Thanks to this design flaw, says Steinsbekk, complementary therapies are not tested for what patients use them for. They are tested instead for the way patients use doctors, and that can be very different, as some of his data suggests.

Steinsbekk has carried out surveys of patients' wants and expectations of complementary medicine – why, for instance, do parents take their children to homeopaths? In Norway, one in four of those visiting homeopaths are children under 10. The commonest complaints are skin conditions and upper respiratory tract infections – coughs, sneezes, earache and the like. Steinsbekk discovered an interesting trend amongst parents. If they were unsure what was wrong with their child they took them to an orthodox doctor. If they knew, or thought they knew, what the problem was, they took them to a homeopath. Clearly parents were not using orthodox doctors and homeopaths in the same way, but were consulting the latter 'as a supplement', says Steinsbekk. He also found that parents seem to choose homeopathy on the basis of recommendations from friends and relatives. Some took their children along despite being sceptical. They did not even need to believe in the treatment, it appeared, being more swayed by others' reports than their own convictions.

An important consideration for Steinsbekk is whether people are wasting their time and money going to see a complementary practitioner. It is only possible to assess this, says Steinsbekk, if they are considered in the context in which the patient pays for them – another reason to ensure that any research is conducted with a view to what the patient, rather than the researcher, wants.

But there are those who feel that biomedical research threatens the very existence of some complementary therapies and could itself harm patients. Paul Dieppe of the University of Bristol outlines a worst-case scenario. A series of clinical trials of, say, a homeopathic remedy, consistently turns up no evidence that it is better than placebo. As a result it is banned. This has a knock-on effect within the complementary medicine community and, for example, a herbalist becomes alarmed that their treatments are also under threat. Consequently, the herbalist adds an ingredient, perhaps an unlicensed use of a steroid, into the remedy that makes it appear to work but dangerous. This is not a completely fanciful idea. There are well-documented cases of apparently very effective herbal remedies containing high levels of strong steroids. Therefore, Dieppe argues, the focus of research into complementary therapies should be on safety, not efficacy. Steps should be taken to ensure that complementary therapists do not do damage by, for example, encouraging cancer patients to forsake conventional treatments.

This does not mean that Dieppe believes research should be abandoned. Far from it – his team is running several studies exploring outcomes and the context and placebo effects. His real concern is that 'CAM practitioners are allowing biomedical researchers to prove that what they do is a waste of time, even when it is obviously valuable'. In doing so, complementary medicine risks being consumed by the medical profession and losing its identity and unique mode of care. It is indisputable that many people, particularly with chronic conditions, experience great benefit from complementary therapies. No amount of research is going to change those individual experiences, argues Dieppe: 'leave them to care and allow people to be cared for'.

○

So what do trials of complementary therapies look like at the moment? Nurse Jenny Gordon is a Research Training Fellow at Napier University in Edinburgh, Scotland. She has an interest in childhood constipation, a condition she became aware of while working in a surgical ward in the late 1990s. Children with intractable constipation would return to the clinic regularly in great distress and be referred to a senior doctor who would treat them according to their expertise. A physician would prescribe laxatives, or a surgeon might operate to remove impacted faeces. Gordon was troubled that the children were receiving crisis management rather than help to overcome the problem. Constipation might appear quite straightforward, but it can be a highly complex problem, which is what makes it very difficult to treat.

A bit of digging convinced her that childhood constipation was generally poorly managed and little researched. There were very few clinical trials into how it might be tackled and scant evidence-based medicine for its treatment. Indeed, doctors couldn't even agree on a clear definition of what symptoms a child with chronic constipation shows. Concerned, she and her colleagues began to consider alternative, less traumatic and more systemic ways of treating constipated children.

As well as being a nurse Jenny Gordon is also a trained reflexologist and was aware that many patients noticed increased urine production and bowel movements after a reflexology session. She decided to see if she could turn this side-effect into a treatment. She ran a small pilot study of 70 constipated children. Those who received reflexology seemed to fare dramatically better than those that did not. However, it was a pragmatic study with poor controls and no randomization – children were given the treatment according to which ward they were on – so Gordon and her colleagues could draw no hard conclusions. The only thing they could say was that it warranted further investigation.

One of the puzzling things Gordon noticed was that many constipated children did not take their prescribed medications. This angered the doctors, who felt their time was being wasted: why would children fail to take a medicine that could ease their distress? The constipation had a huge impact on the children's lives: they couldn't stay overnight with friends, go away to camp or take part in many normal childhood activities. From her experience as a paediatric nurse, Gordon knew that treating children is frequently complicated by the attitudes and beliefs of the rest of the family. Paediatric nurses, like complementary therapists, are used to working in ways that take into consideration all aspects of children's lives, social, emotional, spiritual and so on. She suspected the children were not complying with treatment partly because of the complexities of family life. She had heard, for example, of a grandmother advising against taking a laxative because it 'makes your bowel lazy'.

Gordon decided that before she could start a full-scale randomized controlled trial she needed to find out more about the child's point of view. Why were some resistant to taking medicine? What other factors might be influencing the way they responded? She designed a qualitative study based on interviews and focus groups and put it to the hospital ethics committee. They rejected it. She reworked it and tried again.

They rejected it again. No matter what she did she could not get ethics committee approval for this preliminary qualitative research. The ethics committee felt that there were special circumstances: they were particularly concerned about the implications of interviewing children about their toilet habits. Gordon, an experienced paediatric nurse, could not allay their fears.

By this stage time was getting short and the team decided to go ahead with a randomized controlled trial for reflexology and childhood constipation. They drew up a proposal, put it to the ethics committee who approved it without a murmur.

The study is dividing children with childhood constipation at random into three groups. One will continue with normal treatment, one will receive foot massage but not reflexology and one will receive proper, simple reflexology. The aspect that Gordon is most anxious about is that the massage and reflexology will be administered by the children's parents, not a qualified therapist. The researchers will teach the parents how to do a simple reflexology session, and ask them to follow a particular pattern of sessions at home. The investigators have no control over whether the parents administer the procedure correctly and at the right time – or at all, in fact. However, Gordon hopes the trial will help the parents and children take responsibility for treatment and, pragmatically, it means that her team can see far more children. Parents will receive support but won't need to make a weekly journey to the clinic.

Despite this concern, Gordon believes the results will be significant as the trial is double blind and properly randomized. She is adamant that randomized controlled trials are central to the investigation of complementary therapies, but they may not look quite the same as mainstream ones. The point, she says, is to recognize these differences, understand them and stand up and explain very clearly exactly what each element of the trial is for and why it is being done in that way.

At the time of writing Gordon and her colleagues have recruited half the 180 children they want and results are starting to come in. Since Gordon is blind to which group is which she couldn't tell me what the preliminary results were even if she had wanted to. She is also gathering some qualitative data about the children's situations – not as much as she would have liked, but enough, she hopes, to set the results of the trial in a wider context.

That this study is attempting to measure a more natural situation – parents giving their children reflexology at home – addresses one of the common concerns about clinical trials. They measure, many contend, what would go on in an ideal situation, not what happens out in the real world. This is true for all types of medicine, but once again it can have particular relevance to complementary therapies. An illustration of this was conducted by George Lewith (he of the previously mentioned pelvic pain trial).

Lewith and his colleagues wanted to compare the placebo effect in a clinical trial with what might occur in a more realistic situation, such as a doctor's surgery. They divided people with chronic fatigue at random into two groups and gave one a placebo and the other a supplement designed for chronic fatigue. They then asked the patients which group they suspected they were in. Ninety per cent replied: the placebo group.

The researchers then strove to match more closely the real-world situation, with an 'open label' trial. The supplement was given in its original packaging. All the participants in this round showed significant improvements even though the supplement was exactly the same as the one they got before in an unmarked bottle. In short, had this supplement been tested in a randomized controlled trial it would have come out negative. Yet a patient prescribed it or buying it from their own money might well experience a benefit.

Because of these layers of influence, says Lewith, he no longer conducts quantitative trials in isolation: every one he proposes includes qualitative data collection exploring the range of influences on why patients did or did not get better.

Studies that include an element of qualitative research are becoming more common in orthodox as well as complementary medicines. The *meta*Register of Controlled Trials currently logs 36 such hybrid studies, looking at conditions as diverse as knee pain, epilepsy and hypertension. This is a comparatively new trend, but in time all medical research may be a blend of qualitative and quantitative – who knows?

9

the evidence leads

'A paradigm shift is underway in health care. It will change medical practice in the years ahead', wrote John M. Eisenberg MD, Director of the Agency for Healthcare Research and Quality, the health services research arm of the US Government, in January 2001. Eisenberg was discussing – in the publication *Expert Voices* from the US National Institute for Health Care Management Research and Educational Foundation – the move away from what he called 'hand-me-down' medicine to evidence-based medicine. He did not use the phrase 'paradigm shift' lightly, taking pains to reference its original use by the philosopher of science Thomas Kuhn apropos a revolution in scientific thought.

Evidence-based medicine was first named in 1992 by researchers at McMaster University in Canada. It is a process by which doctors seek and apply the best treatment for their patients. At first glance evidence-based medicine sounds like little more than formalized common sense. Every doctor wants to make the right decisions for their patients; surely only a negligent practitioner would knowingly do otherwise?

A much-quoted *British Medical Journal* editorial published in January 1996 outlined some of the key differences between evidence-based medicine and doctors simply applying the best of their knowledge. Evidence-based medicine is, said the article, 'the conscientious, explicit, and judicious use of current best evidence in making decisions about the care of individual patients'. That is, not just the evidence an individual doctor has acquired during a lifetime's practice – one person's experience – but the best evidence available from every source, every study, every clinical trial. This does not mean that doctors should disregard their clinical experience to become medical automatons. As the article goes on, the discipline: '...integrates the best external evidence with individual clinical expertise and patients' choice, it cannot result in slavish, cookbook approaches to individual patient care'. Clinicians are being exhorted to seek out the best possible evidence gathered by the medical world at large and then apply it to the patients they have before them.

Alongside evidence-based medicine is another important movement, evidence-based health care. This is about making the best decisions for groups of patients: whether to offer a new drug for a particular condition, open a new clinic in a particular region or remove a certain treatment from those available under an insurance scheme, for instance. As evidence-based medicine becomes the dominant way to treat individuals, so evidence-based health care is becoming the preferred way of making large-scale decisions – in other words, how much money is spent and on what.

'Evidence' refers to information that has been obtained by some form of medical research and then published in a medical journal. All the clinical trials discussed in this book will, if and when they are published, form part of this pantheon. The amount of evidence available is enormous and growing. The US National Library of Medicines runs a free electronic database, PubMed, that anyone can access online. This contains over 14 million biomedical publications stretching back 50 years, any one of which might contain information relevant to a patient sitting in surgery.

So let's imagine a doctor, a family or general practitioner, who has just seen a patient with symptoms of bloating, cramps and diarrhoea that suggest a diagnosis of irritable bowel syndrome. In a typical publicly funded clinic she has around 10 minutes to come to this conclusion. After the consultation and following the doctrine of evidence-based medicine she decides to look up the latest research into the condition. The place to look is the published, peer-reviewed literature, and the most straightforward way to do so is to use PubMed, rather like a medical Google. A quick search on the words 'irritable bowel syndrome' produces thousands of hits. Adding the word 'trial' reduces the number of hits to hundreds and the further addition of 'randomized' brings it down to tens. This is still far too many papers to read. A quick trawl of the titles is bewildering: 'A randomized, controlled exploratory study of clonidine in diarrhea-predominant irritable bowel syndrome' or 'Antide-pressants in IBS: are we deluding ourselves?' or even 'Treat-ment of irritable bowel syndrome with herbal preparations: results of a double-blind, randomized, placebo-controlled, multi-centre trial'.

And PubMed is not the only place to search for information. There is a register of current and recently completed clinical trials in the USA alone that produces yet more hundreds of hits on irritable bowel syndrome. The conscientious doctor is stag-gering under the weight of information – it would be totally impossible for her to follow up all these leads. Consider, too, that this is just one of her patients. Every single patient has the potential to call up an equally large amount of data that, in theory, must be trawled through to live up to the aspirations of evidence-based medicine.

This information overload is mirrored by anyone seeking something from the Internet. Much is made of today's 'infor-mation society'. Vast quantities of facts, fiction, opinion and frank lunacy are available on the World Wide Web, and debate rages over whether or not this increases our ability to under-stand ourselves and our world. Some argue that the more information available, the more informed the human race becomes; others urge that raw information is useless unless analyzed and validated.

The situation we have today has both of these ideas running in parallel. Huge amounts of raw data are published on the Internet for all to see – you can go and read the entire three billion letters of the human genome if you so wish. Neverthe-less, some of the most heavily trafficked web sites are those that practice traditional journalism. The two most popular classes of web site – after pornography – are search engines and news sites. Therefore the solution to information overload on the Internet is to get someone to read it for you and produce a précis, which is essentially what good journalism should do. The solution to information overload in medicine is broadly similar: clinical trials are brought together and assessed in what are known as systematic reviews.

One of the first systematic reviews was 'A treatise of the scurvy. In three parts. Containing an inquiry into the nature, causes and cure, of that disease. Together with a critical and chronological view of what has been published on the subject'. It was published in Edinburgh by James Lind in 1753. Lind, the son of a Scots merchant, joined the navy and was promoted to Ship's Surgeon aboard the 50-gun destroyer *HMS Salisbury*. During a voyage in 1747 Lind conducted a clinical trial, though it wasn't called that at the time, on six different treatments for scurvy, the Vitamin C deficiency endemic in the long-distance sailors of the day. The treatments he assessed were cider, elixir of vitriol (a mixture of sulphuric acid, ginger and alcohol), vinegar, sea water, oranges and lemons, and a purgative. He concluded that oranges and lemons were best.

A year later he retired from the Navy and settled down to practice medicine in his native Edinburgh. Still interested in scurvy he gathered up the few other studies that had been done at the time and brought them together in his treatise. This is the essence of a systematic review. Data from clinical trials are considered side by side and a reasonable conclusion is drawn from them if at all possible. Just as journalism is a digest of current affairs evidence, so a systematic review is a digest of clinical evidence. There is one important difference: systematic reviews are conducted transparently. Decisions about which information is included and the significance it is given are set out in the review.

Lind's research eventually led to ships carrying stores of citrus fruits on long journeys and deaths due to scurvy dwindled to virtually zero. Tellingly, it took nearly 50 years from the publication of his first book for Lind's ideas to become universally accepted. Part of the *raison d'être* of evidence-based medicine is to translate proof of principle as quickly as possible into practice.

After Lind's work there were a few other references to the need for data round-ups to draw useful conclusions. A gentleman farmer, Arthur Young, wrote in 1770 that it was impossible to come to a decision about the relative merits of different agricultural methods if they had just been tested with single experiments on different pieces of land. A century later, the Cambridge physicist Lord Rayleigh berated scientists for not recognizing that data gathered by research needs to be evaluated, not just collected.

Systematic reviews are detailed pieces of analysis and require considerable technical skill to produce. The first stage in the process involves searching the medical literature for suitable clinical trials to include in the review. Today this involves trawling electronic databases of publications and identifying any trial that might be of value to the question the review is seeking to answer. The next stage is to make a detailed examination of the relevant trials to assess their validity. This includes taking careful note of how many patients were involved. Was it double blind, single blind or open? Was it placebo-controlled or a comparison of different treatments? How marked were the effects? How reliable were the statistics? Each trial is then given a weighting, and some with no relevance or poor quality data or bad experimental design are discarded completely. More weight, for example, might be given to the results of a trial on 1,000 individuals with double blind controls than one on 10 patients in with no controls. Putting together a systematic review is not for the fainthearted: the Cochrane Collaboration how-to handbook for reviewers runs to 256 pages.

Reviewers must have a profound understanding of clinical trial methodology and considerable statistical expertise. Most doctors are not particularly well versed in either. The person credited with doing most to advance systematic reviews is the epidemiologist Archie Cochrane (1909–1988). Born in Scotland and educated at Cambridge and London Universities, Cochrane went to fight in the International Brigade in the Spanish Civil War when newly qualified in medicine. Following that he served in the Royal Army Medical Corps in the Second World War, during which he was captured and spent time as a prisoner. After 1945 he did a postgraduate epidemiological study of tuberculosis in the USA and spent the rest of his long career in Cardiff, Wales.

In 1972 Cochrane published *Effectiveness and Efficiency: Random Reflections on Health Services*, now considered the original textbook on evidence-based medicine. In it he outlined his ideas on the importance of doing properly controlled randomized trials and then integrating the results into systematic reviews. Shortly afterwards Cochrane's ideas bore fruit in the form of a series of trials and reviews of medicine as practised on newborns. This led to the establishment of the National Perinatal Epidemiology Unit, based in Oxford in the UK and funded by the World Health Organization and the UK Government. In 1985 a team of 50 volunteers, led by Cardiff obstetrician Iain Chalmers, who knew Cochrane and was impressed by his ideas, published a bibliography of 3,500 reports of controlled trials in perinatal medicine. It was a mammoth effort and a convincing proof of principle. At the same time Chalmers established an international collaboration to evaluate health care in newborns. Seven years later the Director of Research and Development in the British National Health Service, approved funding for a research centre 'to facilitate the preparation of systematic reviews of randomized controlled trials of health care' and the first Cochrane Centre opened in Oxford. A year later Iain Chalmers and about 70 other researchers around the world launched the Cochrane Collaboration, 'to prepare, maintain, and disseminate systematic reviews of the effects of health care interventions'. Today this international effort is recognized as the world leader in evidence-based medicine.

At the heart of the Collaboration's work are the Cochrane Reviews. These substantial documents are published quarterly in print and online. They cost between £150 and £350 to access in full, but synopses are available for all to search on the web. Each review provides a snapshot of the latest research into a particular treatment for a specific condition.

This is the answer that has emerged to the problem of information overload. Our imaginary doctor vainly trying to establish the best treatment for irritable bowel syndrome would, in practise, probably do a quick search of Cochrane Reviews. This turns up fewer than 10 hits for the condition, each a paragraph summary. So, with one minute's searching and five minutes' reading, the doctor can be reasonably sure of tapping into most of the latest information on the topic. Cochrane Reviews do not cover every subject, but the library is constantly being augmented and each review is updated on a regular basis, normally every two years or so. If a Cochrane Review exists it is seen as one of the most reliable sources of evidence-based medicine currently possible.

Fifty international expert groups currently contribute to the Cochrane Library. Together they produce hundreds of reviews each year, coordinated by the group editorial teams. As with all scientific publications these reviews are themselves peer-reviewed before acceptance and there is a formal mechanism by which interested researchers can comment on their conclusions.

There are other sources of systematic reviews, though few as comprehensive as the Cochrane Library. Web journal *Bandolier*, run by an independent group of scientists based in Oxford, England, is one widely respected fount of evidence based wisdom. Another is *Clinical Evidence* from the BMJ Publishing group. And increasingly medical journals such as the *New Eng-land Journal of Medicine*, *The Lancet*, the *Journal of the American Medical Association* and the *British Medical Journal* are publishing regular systematic reviews. The evidence base for evidence-based medicine is mushrooming.

But evidence-based medicine, like clinical trials, has nothing to say about how – physically, chemically, biologically – a particular treatment works. It simply focuses on whether and to what extent it is effective and in what individual patient circumstances. As such it is blind to many of arguments surrounding complementary and alternative medicines. If a therapy works – as assessed by clinical trials – from surgery to shiatsu, evidence-based medicine will welcome it into the fold; if it does not it will be rejected.

Many complementary researchers and practitioners are highly sceptical of evidence-based medicine. They feel that it is medicine by rote. Here's a patient showing symptoms x, y and z; enter these into the evidence-based flow chart and out will come the perfect treatment. Evidence-based medicine fans reject this criticism as a caricature, but the idea persists – partly because of the gulf in perspective, or at least the perceived gulf, between complementary practitioners and medical orthodoxy. To most complementary therapists the relationship with the patient is a highly interactive, personal and essential part of the healing process. The training and practice of many complementary therapies emphasizes that no two patients are the same and great weight is laid on the practitioner's intuition. Faced with a treatment regime – evidence-based medicine – that appears to emphasize external information sources over the therapist's expertise, complementary practitioners are often instinctively hostile.

A good way to understand what evidence-based medicine means to doctors and patients is to look at how it is taught. One of the standard evidence-based medicine textbooks for doctors, *Evidence-Based Medicine – How to Practice and Teach EBM*, edited by David Sackett, breaks the process down into five steps.

Step one involves forming answerable questions to the problems the patient is presenting. A question such as 'Are antibiotics the best treatment for my 87 year old female patient's sore throat?' is much easier to answer than 'What is wrong with this patient?'. Detectives, investigative accountants and management consultants take a similar approach.

Step two is to track down the best evidence to answer step one. Doctors are urged to search through medical journals, textbooks and, increasingly, online databases.

Step three is to critically appraise the evidence information collected. Is it relevant to the condition the patient is presenting? Is the evidence applicable to that problem and how big an impact does it have?

Step four is to combine the relevant evidence as sifted out in step three with the doctors' own clinical expertise and the individual biology and circumstances of the patient.

Finally, step five: self-appraisal. Doctors should, according to this rubric, reflect on how well they performed steps one to four and how they might do better in future.

A doctor following these five steps, the reasoning goes, will give his or her patient the best therapy currently available, budget permitting. This is a bold claim, and no evidence-based medicine advocate argues that it is achieved all the time. For example, the same textbook accepts that most practising doctors act only on stage two, assessing the evidence, most of the time.

In fact, the differences between the practice of evidence-based medicine and complementary therapies are not as great as they might appear. At best, both blend general principles with the individual patient's circumstances and the clinical expertise of the doctor or therapist. Both seek the best possible outcome from the available data and both seek to improve as more information becomes available. For doctors that might be another systematic review and for an alternative therapist it may be another practitioner's intuition, but the aim is the same.

○

Even the most ardent supporters of evidence-based medicine accept that it has its drawbacks. Many of these are described in a paper by William Rosenberg and Anna Donald published in April 1995 in *The British Medical Journal*. Evidence-based medicine, they point out, takes time to learn and to practise. As a result, senior doctors running clinics need to develop good management skills to ensure that juniors in their charge have the time to acquire and apply the evidence. Doctors need to become skilled at searching computer databases, which can be a challenge for older practitioners less familiar with the technology. Gaps in evidence, while important for driving future research, need skill to navigate. Finally, suggest Rosenberg and Donald, authoritarian clinicians can see evidence-based medicine as a threat. 'It may cause them to lose face by sometimes exposing their current practice as obsolete or occasionally even dangerous.' That anyone with an Internet connection and some understanding of the terminology can access the same information as the most senior hospital specialist has the potential to demystify the practice of medicine. It's tantalizingly egalitarian.

Small wonder, then, that evidence-based medicine has been given a mixed reception by those who have to put it into practice: doctors. A telling study of 15 Canadian family physicians (GPs) was published in May 2003 in the journal *BMC Family*

Practice. Using semi-structured interviews the authors elicited these physicians' multi-faceted views of evidence-based medicine, then developed a qualitative analysis of those interviews.

The family physicians gave a guarded welcome to evidence-based medicine, admitting that it had improved doctor–patient communication and standards of care, but they were worried about several other consequences. Its lack of emphasis on intuition downplayed something they valued highly. The doctors also felt that evidence-based medicine devalued 'creative problem solving in family practice' and 'the art of medicine' and that guidelines were 'a constraining force on family physicians'. All 15 reported having run into conflicts as they attempted to put evidence-based medicine into practice. Patients' preferences could often be at odds with the evidence, for example. As one physician observed: 'Sometimes it's hard to sell it [the evidence] to certain patients. They have a certain expectation and family medicine is to be patient-centred'. Another admitted that although they explained the evidence to patients to the best of their ability 'we end up doing what the patient wants most of the time'. The researchers conclude by applauding any revisions to the practice of evidence-based medicine that places a greater emphasis on clinical expertise and patient preferences. This, they had shown 'can serve as trumps to research evidence'. (Ironically, these are the very themes that complementary therapists set such store by.)

This chimes with the experience of general practitioner Graham Ward, a doctor of 15 years standing who works in a mixed practice in Bristol in the West of England. He agrees that evidence is an important part of his practice but that it can be a hindrance as well as a help.

Ward worries that some elements of evidence-based medicine have been seized on by governments as a way of measuring doctor performance. This has become particularly significant in the UK, where the incumbent administration has imposed targets for many publicly funded groups, including doctors. Ward says that he has 'certain boxes to tick' founded on an evidence-based approach and gave me an example he had experienced that day.

A patient had come to Ward's clinic complaining of agonizing headaches. On inspection of their notes Ward found mention of an epileptic fit the patient had experienced 10 years previously. The patient had not suffered one since and, according to Ward's judgement, this was not likely to be related to the headaches that were currently being experienced. Nevertheless, he had to question the patient about the epileptic episode and noted that he had done so. It was, he said, probably totally irrelevant to the patient's condition and nothing to do with the reason they were sitting in his surgery crying with headache pain. Yet it was necessary for Ward to include a discussion on epilepsy to, as he put it, 'satisfy our ringmasters'.

Graham Ward doesn't reject evidence-based medicine. Far from it: he requires a good level of evidence of the effectiveness of a drug before he will prescribe it and is cautious about welcoming complementary therapies. What he expresses is pragmatism. As he is fond of saying, 'there is evidence-based medicine and there's common-sense medicine'. Yet his objections are interesting. Is box-ticking a waste of time? Could what appears to be an irrelevance be an unkind intrusion? Or is it a small price to pay for ensuring that doctors do their work carefully and above all, systematically?

Evidence-based medicine has had a similarly mixed reception amongst complementary therapists. Some view it as yet another mechanistic reductionist approach and therefore unsuited to the patient-oriented, individualistic focus of alternative medicine. Others believe it has the potential to further the integration of complementary and orthodox approaches.

Paul Dieppe, for instance, is sceptical about applying biomedical research techniques to complementary medicine.

The point he makes about clinical trials being poor at measuring individualized treatments has a knock-on effect for evidence-based medicine. If the evidence is not available, he argues, there can be no evidence-based medicine. Inevitably, complementary therapies are going to be under-represented.

○

Evidence-based health care draws upon the same type of data as its medical twin, so the arguments about what counts as evidence and how to gather it are equally applicable. The different is that evidence-based health care guides policy, not individual treatments. Crudely put, it is a way of spending limited funds to achieve maximum benefits.

Broadly there are two models of health care funding: payment at the point of delivery – private medical care – or contribution to a fund that pays out when required. That fund might be a private medical insurance company, a government-sponsored insurance scheme or a state-run health care system funded by taxation. In each case, there is considerable pressure on those responsible for spending to do so wisely and to be accountable for their decisions.

BUPA is a health and care company which began in 1947. It has grown into an international organization with a yearly income of almost £3.5 billion and more than seven and a half million customers in more than 180 countries.

BUPA UK Membership is responsible for deciding which clinicians are eligible to treat UK members. At present that includes four complementary therapies: osteopathy, chiropractic, acupuncture and homeopathy. BUPA UK Membership receives regular requests for others to be added to the list. Issues that need to be explored when reviewing treatments that may be eligible for reimbursement include clinical evidence, cost effectiveness and member demand. The evidence is assessed by a medical team drawing on the familiar sources of systematic reviews of clinical evidence and trials.

BUPA is constantly reviewing the services it offers and the treatments it funds and an evidence base is central to this process. The organization does not fund any research into complementary medicine from its own resources, but that is not necessarily the case elsewhere.

In Germany, public health cover is managed by a group of insurance companies. Until a few years ago a loophole in the system allowed individuals to claim for some one-off treatments, including some complementary therapies. That loophole has now been closed, resulting in many patients asking for alternative medicine to be included under the scheme.

At the time of writing, a series of clinical trials for acupuncture to treat migraine, osteoarthritis and backache are in progress, instigated and paid for by German insurers, in response to consumer demand. If the results are positive then it is likely that acupuncture for these conditions will be made available within the German health care system.

○

One of the undercurrents of this book has been something that is really a political (with a small 'p') question. Namely: what is acceptable medicine – no matter that it is called orthodox, regular or conventional – and what is not acceptable medicine, regardless of whether it is labelled charlatanism, quackery or unorthodox? It is a question about boundaries between the two. Are they shifting, should they be retained and how they should be 'policed'? It is also a question about protecting a vulnerable, credulous public from, at worst, the blandishments of the next Lydia Pinkham (a 19th century Massachusetts woman who produced an alcohol-based 'Vegetable Compound' for the treatment of menstrual pain) peddling trumped up 'wonder cures'.

Conventional medicine is currently acceptable medicine. Perhaps its key feature is that it is *scientific*. That is, it is medicine that is based on predominantly western thought of the past 200 or so years, as distinct from folklore or ancient tradition. It tends to rule out non-western systems of medical practice, such as Ayurveda or Traditional Chinese Medicine.

Western science is based on a particular idea about experimental proof – involving a specific set of theories of knowledge. Evidence counts if it has been collected as a result of a well-designed, properly controlled experiment, the results of which have been published in peer-reviewed journals for other scientists to comment upon. Now, though, research into complementary medicine is not only questioning the way evidence is gathered but also challenging the notion of what counts as evidence.

Social anthropologist Christine Barry raised the question of what is to count as evidence in a paper presented to a conference in July 2003. She took a look at much of what this book has covered, but came at it from a rather different angle. Social anthropology, she argued, is very familiar with many of the ideas that makes complementary medicine distinct.

On the one hand, complementary medicine entails holism, the emphasis on the social interaction between practitioner and client as integral to the therapy, and on the patient's ideas about the meaning of their bodily experience. On the other hand, concepts such as 'transcendent, transformational experiences', 'giving meaning', 'changing lived-body experience' or 'intersubjective consensus' are readily understood in anthropology and there is a tradition within that discipline of offering evidence based on these ideas. What's more, she points out, social anthropology has developed the tools to examine these concepts – ideas that biomedicine largely ignores.

Barry argues that social anthropology has the potential to produce analyses that *can* be counted as evidence about complementary medicine. Anthropology's theory of knowledge – including definitions as to what evidence can look like – is very different and might, therefore, suit complementary medicine better than the theories associated with scientific medicine.

Combining the two will be a big challenge – social anthropologists and clinical researchers speak very different languages. But in the opinion of Barry and others like her, it is a must: the current definition of what counts as evidence is, they say, too narrow. The only way to get a complete description of complementary medicine is to expand the definition of evidence beyond its narrow biomedical confines.

○

The issue of what constitutes evidence is not unique to complementary medicine, as one brief final example suggests. In 1999 the UK government published a White Paper outlining its policy on public health, entitled 'Saving Lives: Our Healthier Nation'. It contained the following paragraph:

> Research plays a major role in helping us understand better the causes of ill health.... Public health research is also important in establishing the effectiveness of health programmes but we need to widen the scope of the methods used beyond the randomised controlled trial. In the past it has been the gold standard for research but it is no

longer applicable to all the kinds of research questions which need to be answered.

The sort of public health programmes this paper referred to include opening new hospitals or urban regeneration. Implemented in schools, neighbourhoods or workplaces, these large-scale projects are inseparable from the social interactions. Health and safety at work, for example, rely on employers and employees cooperating. Traffic calming schemes need the support of the local community. These programmes take years to implement and even longer to see results from.

This is all sounding very familiar: randomized controlled trials are not applicable, and there are long-term effects, social interactions and complex interventions. The challenges facing public health planners have many parallels with those facing complementary and alternative medicine researchers. Once again, attempts to get a handle on complementary medicine have potential applications further afield.

10

conclusion

No single technique is emerging to answer the question: 'Are complementary therapies effective?'. What's more, if answers are to come they will likely do so from a combination of different approaches. There is still a debate around the questions 'What does effective mean?' and 'What should be measured and with what technique?'. Research on placebo effects is muddying the water of whether a placebo-controlled trial can provide clear answers, and anyway many of the therapies under discussion do not yet have a suitable placebo. Comparing complementary and orthodox therapies is another option, but they may be working in different ways towards different outcomes – allopathic medicine aiming for a cure, complementary medicine for well-being.

The problems of researching complementary therapies are neither new nor unique. Designing good clinical trials has always been difficult; if it were straightforward the medical literature would not be peppered with arguments over methodology and interpretation of results.

The complexity of the doctor–patient relationship is a challenge to scientific method that is best at examining one element of a system whilst keeping others constant. Yet science deals with complex systems daily.

The big surprise of the Human Genome Project was that we have only around 33,000 genes – way below the 100,000 that many predicted were needed to describe a human being. This dramatically confirmed that the way in which genes work has to be more complex than one gene controlling one function.

In fact, the modern view of how the information in genes controls growth and development is fantastically complex. Switching on a single gene might require anything up to 10 other genes to be active at the same time. And the result of turning on that single gene might influence a whole range of others, turning some on and some off, or changing the rate at which they are read. If one gene has one function, 33,000 genes can only perform 33,000 functions. If one gene works in concert with one or more other genes then the number of possible outcomes rapidly gets into billions.

Geneticists do not, though, throw up their hands and bleat that it is all too hard to study. Instead they design experiments with as few variables as possible, ideally just one, and see what happens. Then they stand back and attempt to fit their new morsel of data into the bigger, more complex, picture. In effect, they switch between viewing the cell as a whole and studying its components in as much isolation as possible. This is, perhaps, not ideal and there will inevitably be some incorrect assumptions made, but it is a way of applying a reductionist method to a complex system.

There are plenty of other examples of this big picture/little picture tactic. Meteorologists do not study the entire global weather system in one go: they break it down into manageable chunks. Ecologists do not assess the impact of global climate change on every shrew and blade of grass: they make generalizations.

It is too glib to say these approaches might be directly applicable to the study of complex interventions in medicine, but lessons could – should – be learnt from studying how science handles complexity in other situations.

○

The whole debate about complementary medicine throws into relief a much bigger question about the medicine we use in general. Many things that doctors currently give their patients have not been through clinical trials. The antibiotic ciprofloxacine has never been tested in children because in animal studies it appeared to cause cartilage damage in juveniles. Nevertheless, it is used with no ill effect when children have an infection that cannot be tackled by another antibiotic. All the empirical – case by case – evidence is that ciprofloxacine is safe; still, this usage has not formally been licensed. Likewise, the treatment of back pain with low doses of the anti-depressant Amitriptyline is highly unlikely to cause a problem, but any doctor prescribing the drug for back pain rather than depression takes a risk. If something goes wrong he does not have any legal support for prescribing 'off licence'.

This is not to say that tests on complementary medicines should be waived. Rather, it means that rejecting complementary therapies purely on the grounds that they have not been fully tested is inconsistent.

○

Chronic conditions and slow degenerative diseases are on the increase. With no cures in sight, palliative care is becoming a bigger part of medicine. Complementary practitioners appear to be comfortable with helping people live with disease rather than attempting to cure it. Many I spoke to talked of the pleasure they get from enabling someone to get on with their lives a little better even though they both know that the problem will never go away. It may be that this success is no more or less than a placebo or context effect. Nevertheless, evidence from psychoneuroimmunology suggests a very real connection between the mind and the body through which these effects might work.

For people like me, with a grounding in biological science, the demonstration of a biochemical link is reassuring. It may have no impact at all on how effective a placebo or context might be, but it fits more comfortably into the biomedical model. And yet it is important to remember that the usefulness of a treatment is totally independent of how it works. Complementary medicine's apparent ability to elicit a powerful placebo effect is the point, rather than how it does it.

That said, access to any sort of health care provision is becoming increasingly evidence-based. If complementary medicine is to feature in the publicly or insurance company-funded system – and there clearly is a demand for it – then it will come under even closer scrutiny. The challenge for complementary advocates is, perhaps, to convince those who hold the purse strings that evidence can come from a wider range of sources than a narrowly defined randomized controlled trial.

This raises a particular paradox for those calling for a plurality in health care, arguing that the complementary approach should live alongside orthodox medicine without needing to conform to its measures of success. Currently, true plurality is only available to those who can afford to pay.

There is a case to be made that randomized controlled trials based on biomedical outcomes do not capture all the elements of many complementary therapies. Adaptations of those trials might get closer to a fuller explanation of what is going on. Qualitative research may well help to develop those tweaks and provide valuable insights unconnected to the effectiveness of a treatment.

There are some intriguing and currently inexplicable findings surrounding complementary therapies. Why, for example, does the Bowen Technique appear to work so well for frozen shoulder? What is the explanation behind the apparent ability of homeopathy to be more than a placebo? These results present some very profound challenges to science's explanation of how our bodies work. This can only be a good thing. The scientific method is a powerful tool for investigating the unknown, and research thrives on puzzles. Attempting to answer these questions will certainly produce new data. It may be as mundane as spotting the hidden flaw in the research, or it may uncover a previously unknown function of our bodies.

Pursuing these puzzles has another potential benefit. Testing complementary medicine is difficult. In science, doing difficult things often results in better tools. Even if every single complementary therapy turns out to have a simple explanation, extracting it is going to hone medical research. That in turn will improve the way in which orthodox medicine can be studied.

It is appropriate to end with a conclusion drawn by someone who has spent his working life as one of the most senior and respected medical researchers, and who has been closely associated with the development of randomized controlled trials and with the establishment of the Cochrane centre:

> the most important resource required to promote the concept of integrated health care is likely to be humility among those whose practices will be put to the test, within both orthodox and complementary medicine.

Professor Sir Iain Chalmers, 1998

further reading

Australian Government Expert Committee on Complementary Medicines in the Health System Report to the Parliamentary Secretary to the Minister for Health and Ageing (2003) *Complementary Medicines in the Australian Health System*, September. Available at http://www.tga.gov.au/docs/html/cmreport1.htm.

Bandolier: http://www.jr2.ox.ac.uk/bandolier/.

Barnes, P., Powell-Griner, E., McFann, K. and Nahin, R. (2004) *Complementary and Alternative Medicine Use Among Adults: United States, 2002*. CDC Advance Data Report #343, 27 May.

Berman, J.D. and Straus, S.E. (2004) Implementing a research agenda for complementary and alternative medicine. *Annual Reviews of Medicine*, **55**, 239–54.

BMA News (2004) Complements of the house. London: British Medical Association, 22 May.

British Medical Journal (2004) Clinical Evidence: the international source of the best available evidence for effective health care. British Medical Journal Publishing Group Ltd. Available at http://www.clinicalevidence.com/.

Cochrane Reviews: http://www.cochrane.org/.

Cohen, I.R. (2000) *Tending Adam's Garden*. Academic Press: San Diego, CA.

Coward, R. (1989) *The Whole Truth: the Myth of Alternative Health*. London: Faber.

Department of Complementary Medicine, Peninsula Medical School (2004) *Complementary Medicine: The Evidence So Far.* A documentation of research 1993–2003. A summary of our most important research findings to date. Exeter: Universities of Exeter and Plymouth.

Earl-Slater, A. (2002) *The Handbook of Clinical Trials and Other Research* Abingdon, Oxfordshire: Radcliffe Medical Press.

Evans, D. (2003) *Placebo: the Belief Effect*. London: HarperCollins.

Evans, P., Hucklebridge, F. and Clow, A. (2000) *Mind, Immunity and Health: the Science of Psychoneuroimmunology*. London: Free Association Books.

Freidson, E. (1980) *Patients' Views of Medical Practice: A Study of Subscribers to a Prepaid Medical Plan in the Bronx*. Chicago: University of Chicago Press.

Ga²len (2004) *Global Allergy and Asthma European Network: Network of Excellence*. European Union, 6th Framework Programme. Available at http://www.ga2len.com/.

166 the whole story

166 the whole story

166 the whole story

166 the whole story

166 the whole story

Global Polio Eradication Initiative (2003) *Strategic Plan 2004–2008*. World Health Organization Publications. Available at http://www.polioeradication.org/.

Gray, A. (ed.) (1993) *World Health and Disease*. Milton Keynes: Open University Press.

Muir Gray, J.A. (2001) *Evidence-Based Healthcare*. London: Churchill Livingstone.

Guess, H.A., Kleinman, A., Kusek, J. and Engel, L. (eds.) (2002) *The Science of the Placebo: Toward an Interdisciplinary Research Agenda*. London: BMJ Books.

House of Lords Select Committee on Science and Technology Sixth Report (2002) *Complementary and Alternative Medicine*. London: The Stationery Office, 21 November. Available at http://www.parliament.the-stationery-office.co.uk/pa/ld199900/ldselect/ldsctech/123/12302.htm.

Illich, I. (1977) *Limits to Medicine: Medical Nemesis – The Expropriation of Health*. London: Penguin Books Ltd.

Jenkins, T., Campbell, A., Cant, S., Hehir, B., Fox, M. and Fitzpatrick, M. (2002) *Alternative Medicine: Should We Swallow It?* Institute of Ideas: Debating Matters. London: Hodder & Stoughton.

Karpf, Anne (1988) *Doctoring the Media: the Reporting of Health and Medicine*. London: Taylor & Francis.

Lewith, G.T. and Aldridge, D. (1993) *Clinical Research Methodology for Complementary Therapies*. Singular Publishing Group.

The James Lind Library: Documenting the evolution of fair tests of medical treatment. Available at http://www.jameslindlibrary.org/.

Mackay, H. and Long, A.F. (2003) The experience and effects of shiatsu: findings from a two country exploratory study. *Report no. 9*. Salford: University of Salford Health Care Practice R&D Unit.

Martin, P. (1998) *The Sickening Mind: Brain, Behaviour, Immunity and Disease*. Flamingo.

Mason, J. (2002) *Qualitative Researching*, 2nd edn. London: Sage.

McKeown, T. (1976) *The Modern Rise of Population*. London: Edward Arnold.

Murphy, E. and Dingwall, R. (2003) *Qualitative Methods and Health Policy Research*. New York: Aldine.

Murphy, E., Dingwall, R., Greatbatch, D., Parker, S. and Watson, P. (1998) Qualitative research methods in health technology assessment: a review of the literature. *Health Technology Assessment*, **2**(16), 1–276.

Peters, D. (ed.) (2001) *Understanding the Placebo Effect in Complementary Medicine*. London: Churchill Livingstone.

Porter, R. (2002) *Blood and Guts: a Short History of Medicine*. London: Allen Lane.

Sackett, D.L., Straus, S.E., Richardson, W.S., Rosenberg, W. and Haynes, R.B. (2000) *Evidence-Based Medicine: How to Practice and Teach EBM*, 2nd edn. London: Churchill Livingstone.

Seale, C. (2002) *Media and Health*. London: Sage.

Sharma, U. (1992)*Complementary Medicine Today: Practitioner and Patients*. London: Routledge

World Health Organization (2001) *Legal Status of Traditional Medicine and Complementary/Alternative Medicine: A Worldwide Review*. Available at http://www.who.int/medicines/library/trm/who-edm-trm-2001-2/legalstatus.shtml.

Index

In our rapidly changing world it make
for us to use every advantage we can. For example, no longer is it necessary to retype a letter numerous times to get it perfect—we can perfect it using a word processing program on a computer before it is ever put on paper.

At the computer's heart is a small device known as a chip. This chip is made of silicon, the same substance that is the source for quartz crystals.

It has been discovered that we can make use of crystals on a variety of levels—from the heart of a computer to spiritual healing. Numerous books explain the former; this book explains the latter.

This book looks at the latest discoveries in the world of crystallography. It not only shares them with you, but also *shows you how to make use of them in your day-to-day life.* This includes not only the use of crystals for healing, but also for everything from working with dreams to improving your sex life.

This book is different! Many books discuss the value of raw crystals. These form the first generation of books on the spiritual value of crystals. **CRYSTAL HEALING: THE NEXT STEP** explores newly discovered information on the value of polished and shaped crystal, and shows you how you can use the vastly improved abilities of this type of quartz to improve your life. This is a book you will not want to miss!

About the Author

Born and raised on a farm in North Dakota, Phyllis Galde had ample opportunity to be close to the Earth and become acquainted with the mineral kingdom. Living in a haunted house introduced her to the world of spirits at an early age, and offered her many opportunities to be comfortable with other dimensions. Always a lover and caretaker of animals, she transferred that caring into teaching school for sixteen years. She is an ordained priest with the Order of Melchizedek. She has worked with many New Age organizations, including the Edgar Cayce A.R.E. and Theosophy, and is a MariEl healer. She is currently editor in chief of FATE magazine.

To Write to the Author

If you wish to contact the author or would like more information about this book, please write to the author in care of Llewellyn Worldwide, and we will forward your request. Both the author and publisher appreciate hearing from you and learning of your enjoyment of this book and how it has helped you. Llewellyn Worldwide cannot guarantee that every letter written to the author can be answered, but all will be forwarded. Please write to:

Phyllis Galde
c/o Llewellyn Worldwide
P.O. Box 64383-246, St. Paul, MN 55164-0383, U.S.A.
Please enclose a self-addressed, stamped envelope for reply,
or $1.00 to cover costs.
If outside the U.S.A., enclose international postal reply coupon.

Free Catalog from Llewellyn

For more than 90 years Llewellyn has brought its readers knowledge in the fields of metaphysics and human potential. Learn about the newest books in spiritual guidance, natural healing, astrology, occult philosophy and more. Enjoy book reviews, new age articles, a calendar of events, plus current advertised products and services. To get your free copy of *Llewellyn's New Worlds of Mind and Spirit*, send your name and address to:

Llewellyn's New Worlds of Mind and Spirit
P.O. Box 64383-246, St. Paul, MN 55164-0383, U.S.A.

ABOUT LLEWELLYN'S NEW AGE SERIES

The "New Age"—it's a phrase we use, but what does it mean? Does it mean that we are entering the Aquarian Age? Does it mean that a new Messiah is going to correct all that is wrong and make the Earth into a Garden? Probably not—but the idea of a *major change* is there, combined with awareness that the Earth *can* be a Garden; that war, crime, poverty, disease, etc., are not necessary "evils."

Optimists, dreamers, scientists . . . nearly all of us believe in a "better tomorrow," and that somehow we can do things now that will make for a better future life for ourselves and for coming generations.

In one sense, we all know there's nothing new under the Heavens, and in another sense that every day makes a new world. The difference is in our consciousness. And this is what the New Age is all about: it's a major change in consciousness found within each of us as we learn to bring forth and manifest powers that Humanity has always potentially had.

Evolution moves in "leaps." Individuals struggle to develop talents and powers, and their efforts build a "power bank" in the Collective Unconscious, the "soul" of Humanity that suddenly makes these same talents and powers easier to access for the majority.

You still have to learn the "rules" for developing and applying these powers, but it is more like a "relearning" than a *new* learning, because with the New Age it is as if the basis for these had become genetic.

Other books by Phyllis Galde

The Message of the Crystal Skull
    (with Alice Bryant)

Forthcoming books

The Dream Crystal Book
    (with Frank Dorland)

Crystal Balls: Their Magic and Lore

Skull Magic

Edited books

Holy Ice by Frank Dorland

Llewellyn's New Age Series

CRYSTAL HEALING

THE NEXT STEP

Phyllis Galde

1996
Llewellyn Publications
St. Paul, Minnesota 55164-0383, U.S.A.

FIRST EDITION
Fifth Printing, 1996

Cover Photo: Frank Dorland
Cover Design: Terry Buske
Photos: Frank Dorland
Illustrations: Brooke Luteyn
Technical Consultant: Frank Dorland

Library of Congress Cataloging-in-Publication Data
Galde, Phyllis, 1946-
    Crystal healing.

    (Llewellyn's new age series)
    Bibliography: p.
    1. Quartz crystals—Miscellanea. 2. Healing—Miscellanea. I. Title. II. Series.
BF1442.Q35G35   1988   133.3'22   88-45106
ISBN 0-87542-246-2

Llewellyn Publications
A Division of Llewellyn Worldwide, Ltd.
P.O. 64383, St. Paul, MN 55164-0383

Acknowledgments

A special thank you to all who made this book possible. To my parents, Osborne and Doris Galde, for allowing me the freedom to grow. To my daughter Angie for her assistance. A warm thank you to Frank and Mabel Dorland for their hospitality. Without their years of experience which they so generously shared with me, this book never could have been written. Thank you Thomas Begich for giving me the space to write. Nancy Mostad, a special blessing for all the hours of typing and encouragement. Thanks to Donald Michael Kraig for his expertise. A special thank you to each and everyone at Llewellyn for being patient with me while I was writing this. Thank you Carl and Sandra Weschcke, and Steve Bucher for the opportunity to write this. For your efforts, thank you Terry Buske, Julie Feingold, and Brooke Luteyn. And to all the beautiful crystal beings out there, Love and Light to you all.

CONTENTS

Lives with a Crystal, Crystal Learning, Future Use of Crystals, Crystals for Psychic/Spiritual Development, Myths About Crystals

To everything there is a season, and a time to every purpose under the heaven;

A time to be born, and a time to die; a time to plant, and a time to pluck up that which is planted;

A time to kill, and a time to heal; a time to break down, and a time to build up;

A time to weep, and a time to laugh; a time to mourn, and a time to dance;

A time to cast away stones, and a time to gather stones together; a time to embrace, and a time to refrain from embracing;

A time to get, and a time to lose; a time to keep, and a time to cast away;

A time to rend, and a time to sew; a time to keep silent, and a time to speak;

A time to love, and a time to hate; a time of war, and a time of peace;

Ecclesiastes 3:1-8
Bible, King James Version

Introduction

What *is* the truth about crystal healing? How do you *really* heal with crystals? This book will show you how to heal with crystals in an easy and effective way.

How can you best use crystal power? What value can crystals actually have for you? So many books have been written on this subject, and so many viewpoints have been presented that it's easy to become confused as to what guidelines you should follow. This helpful book will give you some fundamental groundwork on how to go about using crystals and benefiting from them.

Why use quartz crystals? What is it about them above all other stones that makes them so special?

Quartz crystals have been used since ancient times as powerful healing objects, meditation tools, and medicines (in the form of elixirs). Wise adepts have long known about their qualities and have used crystals for powerful talismans and amulets. Throughout history people have valued the beauty of quartz crystals and have used them for ornamental decoration. References to crystals are found in both the Old and New Testament (Ezekiel 1:22, Revelations 21 and 22), and in many other sacred teachings throughout the world.

How can minerals or stones have any influence or value? It is because there is a consciousness inherent in *all* forms of matter. Even rocks have a consciousness, and a crystal is much more than a rock. Rocks are made by pressing mineral matter together, but crystals actually grow, cell by cell, in a perfectly formed and engineered electronic structure (engineered by Mother Nature who also controls the growth). We use these crystals of the mineral kingdom to aid us in our attunement with the different aspects of ourselves.

Crystals have their own particular vibration of a precise and measurable intensity. This vibration attunes itself to our human vibration better than any other gem or mineral. Quartz crystal is used to amplify, clarify, and store mental energy or mental records. When you create a thought, you can amplify and clarify it by using a crystal. Quartz has long been recognized for its ability to produce electrical impulses. Pressure on quartz generates a minute electrical charge called piezo-electricity, which is an important source of energy in this Space Age of technology.

Crystals have long been revered for use in magick, for psychic development, and to see into the "hidden dimensions" that permeate physical reality. It is said that crystal has the ability to rebroadcast energy from the Universal Mind so your "inner self" can pick it up, granting you heightened perception. A crystal is a focus for this knowledge and help, and it magnifies and transmits psychic energies and healing powers.

Crystals have a hexagonal, symmetrical shape. This is a basic form in geometry, physics and atomic theory, and is a universal, perfect form in the structure of matter.

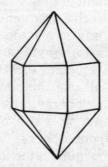

Structure of Quartz Crystal

Although it is your mind that really heals you, quartz crystal amplifies this effect, and enhances your mind control and will power. The mind and other influences can activate the immune system. Recent research indicates that it is possible to increase the interferon levels in the bloodstream by external signals and mental control.

Uses of Crystals

Perhaps the *greatest* value of crystals is their use in healing. They have long been known for their curative effects when used in the preparation of tinctures; for their protective characteristics when worn as amulets and talismans; and for their ability to enhance the energy fields of the body as they emit uniform vibrations in harmony with the natural vibrations of the human body. This is a "boosting" or beneficial effect.

The energy field of quartz and its ability to oscillate at a precise frequency is known as *piezoelectricity,* and is a major contributor to our communications systems. Computers, sonar, watches, medical electronics, TV and radio stations—all use this amazing and constant, undeviating energy to select and isolate a chosen vibration so it can be used without interference from others. "Piezo" means applied pressure-generated electricity, and quartz is the most economical source of this energy. To reduce the cost even further, laboratories are growing quartz from a slice of the real thing. It has been found that the laboratory grown crystals are superior in every way to all but a few rare, almost perfect natural crystals.

The vibrational frequency depends on the crystal shape and size when cut and used in electronic devices. This interesting phenomenon was discovered by Madame Curie and her brothers in their Paris lab in the 1880's.

In the metaphysical approach, it is believed that the crystal reacts differently because its energy is received from the human mind and body, not from an electrical generator or DC battery. One reason why psychic crystals are carved in smoothed and rounded surfaces is because these curved, smooth surfaces help them act as reflecting amplifiers for the mind and body.

Crystals do not actually hold an electrical charge as such (like a battery), but they do have a memory. They act as if they are charged when held in the hand and "turned on" by direct contact with the human body. This is their amplifying ability of the energy received from the body cells which carry an electrical current or charge. Because of this quality, these minerals amplify healing and speed up the healing process. Quartz crystals also enhance the meditative state and allow one to enter an alpha state more easily.

Crystals are excellent conductors of energy. They can even absorb one kind of energy and emit another when squeezed. However, quartz crystal is one of the poorest conductors of heat or cold. Rapid temperature changes crack crystal, as it is almost impossible to contract, expand, or conduct thermal changes. In the metaphysical approach quartz crystal is thought to absorb both magnetism from the Earth's core and radiation from the Sun, and re-emits that energy. Kirlian photography has recorded this energy

emission, which appears in photographs as a
white-light aura radiating from a blue star-like
center. The energy that radiates from clear
quartz and other crystals resonates with the
human aura—this is its healing attribute.

This resonance occurs rapidly, within a few
moments of holding a crystal or gemstone in
the hand. Energy from crystals passes through
and penetrates all bodies of matter, penetrat-
ing even into human cellular structure.
Crystal energy transmission has a magnetic
polarity similar to the natural aura polarity
used in laying on of hands. Its strong ability to
match aura energy and resonate with it makes
crystal a powerful healing tool. Clear quartz
contains the greatest human resonance and
ability to transmit any color, to focus a chosen
strand of its rainbow white light spectrum and
use it to transmit, store, duplicate and magnify color and aura polarity.

An interesting phenomenon happens when
you begin to work with crystals — for whatever reasons. You will start becoming aware of
an energy or force greater than what you
presently contain. This force has been called
your Higher Self, and it encompasses "that
which you are capable of becoming." It is your
perfected self. Quartz crystals, in their won-
derfully helpful way, will help you tune into
this higher aspect of yourself.

Nature has created a beautiful form of a
regular six-sided crystal. Sometimes we have

been able to improve on what Nature has provided us with. She has given us the raw materials and it is up to us to use these materials to make our lives better and easier. For example, we have used iron to make steel, which is a superior alloy. We have used aluminum, glass, lead, zinc, etc., to make automobiles.

So it is with quartz crystal. We have learned how to use this undeviating electrical energy in many forms of technology and communications. In order to use quartz for these devices, it must be cut or sliced in a specific fashion and in a specific size.

We are just beginning to rediscover how to use crystals for healing and divination. This book will show you *how* and *why* a crystal's powers are so greatly enhanced when they are smoothed and polished, and how the use of a rounded, smoothed crystal as a healing or divination tool results in an incredible amplification of one's own natural abilities.

Read on to learn more about the *next step* of crystal healing. You will learn how to handle and work with crystals to heal yourself, others, and the Earth.

Chapter One

HISTORICAL USES

Stories of the powers and uses of crystals and gems have come down to us from the beginning of time. Legends have it that crystal forces "set" the electromagnetic field of the Earth so that human souls could incarnate. The most popular stories dealing with the use of crystals date back to Atlantis. There, it is said, large crystals were used to generate power for cities, and it was the abuse of these energies that eventually caused the destruction of this great civilization. Edgar Cayce (one of America's most famous seers) stated that the largest crystal generator is buried under the Atlantic Ocean near the Devil's Triangle, and this massive shift of unfocused electromagnetic energy is what causes ships and planes to go astray.

It is said that crystals were harnessed in Atlantis for power and surgery. Some have claimed that in ancient Egypt they were the force that enabled the huge sandstone blocks to be positioned in building the pyramids. Built atop granite, the immense pressure of these stones activated the crystals found naturally in this granite, creating a gigantic generator.

Crystals were used to light the inner tunnels and chambers while the pyramids and Mystery Temples were under construction. Ancient civilizations, particularly Atlantis, used giant crystals to focus laser light. Crystals were used to fly aircraft, light homes, heal, for agriculture, to focus beams of energy between pyramids, obelisks, temples, stone monuments and all grid points. Each pyramid amplified the energies to "light" the Earth. Crystals were used to control weather, to attune initiates, and in radio waves for communicating with home bases in space—the orbiting Mother ships.

Crystals were used to generate energy which was focused in various ways, not all of which were positive. The Atlanteans used an advanced form of hypnosis in which complex detailed visuals were projected into a person's brain, either knowingly or not. By doing this, thoughts could be influenced, as could memory banks. Near the decline of Atlantis, the dark priesthood was involved in control over others. They used crystal power to create pestilences and diseases to kill people by projecting holographically the images, fears, concepts, etc. they wanted to impress upon the people. They experimented on the populace. Scientists manipulated embryos to create subhuman forms to be used as slaves.

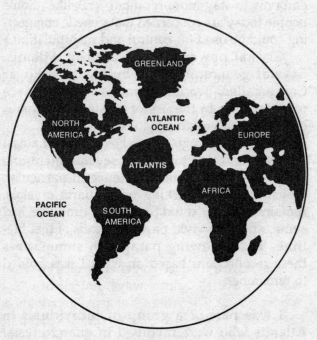

Atlantis

Embryo development was arrested with crystal energy, and hypnotic suggestion caused the embryos to stay more reptilian, scalelike. (Some people today are concerned that genetic engineering could be used for control and manipulation.)

Crystal power was misused in Atlantis. Crystal generators were built which Edgar Cayce called "terrible crystals." Many Cayce readings refer to the use of crystals in Atlantis, both for positive and negative uses.

Several years ago, when I experienced a past life reading, I was told about several interesting lives in Atlantis. One experience in particular, which involved working with quartz crystals. remained in my mind.Since that time I've had some especially vivid past-life recalls of that lifetime. The following paragraph summarizes these recollections based on what I was guided to remember.

I was part of a group of individuals in Atlantis who were involved in quartz crystal experimentation on the sensations and happenings in the physical body. In that lifetime, it was learned that much good could be accomplished in the blood and nervous systems. Unfortunately, the people in higher positions then changed the focus of my duties into a more technical advancement where the work with crystals dealt with the resonating and radiating effects of crystal energy. The effects that the stone magnified were used for dominating a physical body, and I came to feel that this was trespassing on the inherent rights and privacy of an individual. Finally, I was removed entirely from the scientific department in that lifetime.

This book, then, is an attempt to show that crystals should only be used for positive results such as healing with free will, and not to control anyone.

In Atlantis as well as other civilizations, the earth grid system or energy ley lines were understood and utilized. Crystals were employed to accentuate this energy grid system. Crystal energy was used as a focus for other purposes also. The people of Atlantis and Lemuria used ultrasound and other energy forms. They used mind power by humming or "toning," and by so doing levitated rock. Crystal pyramids were used like a laser device because they could store and focus energy, such as sunlight energy. The crystals could encode information to a higher vibration and beam information into a person, so a student could obtain the equivalent of a "college education" from crystal energy.

Although many people are familiar with the lost continent and advanced civilization of Atlantis, fewer have heard of Lemuria, or Mu. Lemuria supposedly was a gigantic continent that predated Atlantis. Part of it was located where Hawaii (some believe it extended to California) is today, and it reached to Mongolia.

Famous Ley Line in England

More paranormal happenings occur along ley lines (the Earth's energy grid system) than anywhere else. Some researchers believe there is an abundance of quartz crystal located underground along these lines, which may partially explain the source of the energy emanating from the ley lines. This is an aerial view of the Saintbury ley in Gloucestershire, England which extends for about 3½ miles NNW-SSE. (The white lines and circles have been drawn in to show where the energy flows are located.)

Easter Island Giants
Were these the unfriendly giants the Lemurians were afraid of?

It is believed by some to be the evolution of the first human in physical, flesh form about 18 million years ago. Lemurians were peaceful, gentle people.

Lemurians used crystals underground to grow food because they were afraid of the giants living on the surface who were unfriendly and antagonistic. The Lemurians came to spend most of their time in caves, and so needed light energy to grow food. The stored energy from crystals was the source for this light. The huge stone relics of giant heads found on Easter Island are thought by some to be fashioned after these giants who dominated the Lemurians.

In times past, crystals were used to balance harmonies in the body — to stabilize the flow of prana (invisible energy in the air, or the vital life force), to stimulate the chakras (spiritual energy centers in the body), to raise the kundalini energy (spiritual energy lying dormant in the body). It is believed that Atlanteans cut quartz crystal in special ways for healing. Facets of three, five, or seven sides were used for certain illnesses, while four- and eight-sided quartz crystals were used to maintain balance. Crystals of specific colors—red, green, blue, violet, white, etc.—were used to heal various illnesses. Their healing properties restored eyesight to some in the hands of adept healers. Crystals were placed on the eyelids to in draw prana and heal organs inside the body.

The ancient Greeks believed that quartz was "eternal ice" which came down from Mount Olympus, home of the Greek gods. The people used its natural magnifying power to focus the heat of the sun's rays in order to start ceremonial fires. According to the famous Roman scholar Pliny, who wrote in the first century A.D., they used the focused crystal heat from the Sun to cauterize wounds.

The ancient Egyptians held quartz sacred, and carved their drinking vessels from quartz. When they drank from these cups, they believed that the water became imbued with life-giving energy. They also mined quartz, and along with many cultures began to carve it into jewelry and a variety of objects both artistic and utilitarian.

American Indian Legend of Crystals

In ancient times, people lived in harmony with Nature. They spoke the same language as the animals and plants. They hunted for food only to satisfy their hunger and needs, always offering a prayer of thanks for what they had taken from Nature.

As time went on, humans lost this innocence and harmony. They took more than they needed. They forgot their prayers of gratitude. They killed animals, and each other, for sport or pleasure.

The Bear Tribe, chief among the animals, called a meeting of all the animals. They decided that something had to be done. The Bears suggested that they shoot back when the humans shot at them, but the bow and arrow required too great a sacrifice, for one bear would have to give up his life so that his sinew could be used for the bowstring. The bear's claws were too long for shooting a bow anyway, and would become entangled on the string.

The Deer Tribe offered another method of dealing with the problem. One of their members said, "We will bring disease into the world. Each of us will be responsible for a different illness. When humans live out of balance with Nature, when they forget to give thanks for their food, they will get sick." And in fact the Deer did invoke rheumatism and arthritis; each animal then decided to invoke a different disease.

The Plant Tribe was more sympathetic and felt that this was too harsh a punishment, so they volunteered their help. They said that for every disease a human gets, one of them would be present to cure it. That way, if people used their intelligence, they would be able to cure their ailments and regain their balance.

All of Nature agreed to this strategy. One plant in particular spoke out. This was Tobacco, the chief of the plants. He said, "I will be the sacred herb. I will not cure any specific disease, but I will help people return to the sacred way of life, provided I am smoked or offered with prayers and ceremony. But if I am misused, if I am merely smoked for pleasure, I will cause cancer, the worst disease of all."

The close friends of the Plant Tribe, the Rock Tribe, and the Mineral Tribe agreed to help. Each mineral would have a spiritual power, a subtle vibration that could be used to regain perfect health. The Ruby, worn as an amulet, would heal the heart; the Emerald would heal the liver and eyes, and so on. The chief of the mineral tribe, Quartz Crystal, was clear, like the light of Creation itself. Quartz put his arms around his brother Tobacco and said, "I will be the sacred mineral. I will heal the mind. I will help human beings see the origin of disease. I will help to bring wisdom and clarity in dreams. And I will record their spiritual history, including our meeting today, so that in the future, if humans gaze into me, they may see their origin and the way of harmony." And so it is today.

This is a Cherokee legend, but it has been told in almost every tribe in the Americas. It tells of an ancient time of peace, a mythical homeland known to every culture on Earth. The Native Americans call it the "old way" or the "original way."

American Indians valued quartz crystal highly, and many accounts of its use occur throughout their history. The Apache Indians claimed that quartz crystal gave them the ability to locate their stolen ponies even over great distances.

Later in history, the American Indian shamans placed quartz crystals over their own eyes to help them become clairvoyant. Indoctrination into shamanism in certain tribes involved placing a tiny seed crystal (small bean or seed size) between the bottom two ribs on the left side of the body. This crystal was removed from its "nest" on the shaman's death and re-inserted into a new young shaman on his "graduation," thus giving him instant all-knowing of his predecessor.

Ancient Gemstones

Most gemstones discovered from early times usually measured one to three inches in length. It did not make a difference to these ancient people if the stones were sapphires, diamonds, etc. Many gemstones which were rare, beautiful and resistant to breaking were special to them.

Craftsmen rounded only the large ones to a cabochon shape.

rounded dome

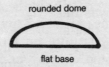

flat base

They would try to save as much of each stone as possible, and just round and polish them. (None of the ancient gems that were found were faceted, as this technique hadn't been discovered yet.)

Diamonds were one of the precious, valued gemstones used by the ancients. The origin of the word diamond is very interesting. Diamond comes from the Latin word *adamas*, which in turn originated from the same Greek word. Adam represented the first archetypal man. The word *adamant* means "a legendary stone believed to be impenetrable," also, "an extremely hard substance." A diamond is hard and unyielding, and mysterious. Early artists would smash all the small diamonds, along with sapphire and any other stones that were clear, only engraving on the large ones. They used the ground up gem dust for polishing, grinding, and engraving.

Many of the precious gems had engraving on the front side. Some of these engraved stones were found on breastplates which were used by high priests such as those mentioned in the Old Testament. Other engraved gemstones were used as seals. Some of these ancient seals which have been found are three or four thousand years old. A tiny, cylindrical seal carved out of quartz crystal was unearthed which was a tax record, showing that an individual had paid his taxes — much like a modern day paper receipt, but far more enduring. Because tax records are one of the more common and numerous artifacts that archaeologists find, it would seem that taxes have been the most important thing throughout history!

The early crystal tax seals were carved on small rollers (an inch or so), like a handleless rolling pin. The tax records were in clay tablets. The crystal seal was rolled over the damp clay, thus giving a bas relief impression. The clay tablets were then baked for permanence.

←1¼″ in length→

Sumerian Tax Seal
Made from Quartz Crystal, c. 3500 B.C.

Quite often the seals would be carved with the likeness of the water god Ea, a water god who had replaced the older queen of heaven. (When patriarchal societies replaced the matriarchies with force, they usually exchanged the original female deities with their choice of a male deity.) Ancient people believed that all of the stars were crystals which the queen of heaven held in her hands. Thus she was known as the patron saint of crystals. One of the first names for this goddess was Asteria, the Starry One, which comes down to us in a different age, several thousand years later, as Astraea, the Roman goddess of justice.

Quartz crystal has had immense value and effect in most cultures, and exists in many universal legends throughout the world. The stories of its powers date far back into prehistoric times. Ancient priests believed quartz was a God-given gift of force which defied all evil. It could dissolve enchantments and spells, and effectively worked against black magic. As a magical talisman, crystal has never been equaled.

Because so many mystics and wise people of every culture revered quartz crystal, it certainly must possess special qualities even though they are unmeasurable by today's scientific standards. Hopefully, in this coming Age of Aquarius we will again learn to utilize the full potential of this wonderful, magical gift from the mineral kingdom.

Chapter Two

ORIGIN OF
QUARTZ CRYSTAL

A crystal is a beautiful, perfect form. It contains within it harmony, balance, clarity and perfection. A naturally occurring quartz crystal may take over a million years to reach its present condition. Quartz comes from deep within the Earth's core, and was formed when the Earth was evolving.

Natural quartz crystals, often referred to by ancient traditions as the "veils of the earth," frozen water, or frozen light, are formed naturally from the elements silicon and water through a lengthy process involving heat and pressure. They are buried in the Earth, or sometimes in streambeds where they have washed down from higher ground after they have been dislodged. They are often found with gold and/or silver. Many other varieties of quartz crystal are also found worldwide, especially where volcanic action has occurred. Large numbers of crystals

15

are now being mined in Arkansas, Brazil, Colombia, Sri Lanka and other mineral-rich countries.

The name crystal comes from the Greek word *krystallos*, meaning "clear ice," for the ancient Greeks thought that these transparent rock crystals were in fact frozen water turned into stone, and that rock crystal was eternally frozen. Another legend has it that Holy Water was poured out of the Heavens by God and frozen to ice in outer space on its voyage to Earth. Angels petrified the "Holy Ice" to preserve it in solid form as a protective blessing for humanity.

Approximately ninety per cent of the Earth's crust is made up of the mineral group known as silicates, a combination of silicon and oxygen, plus other elements. The simplest silicate is merely silicon and oxygen—quartz crystal. Chemically, it is the oxide of the element silicon, and its chemical formula is SiO_2 (silicon dioxide). It has a hardness of 7 on the Mohs scale (see diagram). The crystal structure of quartz is hexagonal with void spaces in geometric trails throughout the crystal. Natural grown crystal looks like it has had its sides faceted, but in fact this is how Nature has marvelously formed it.

Quartz is the most resistant mineral to weathering. It is very durable and is able to scratch steel and glass, yet it can shatter if it's dropped. Its normal appearance is like a six-sided prism. The orderly, consistent pattern of

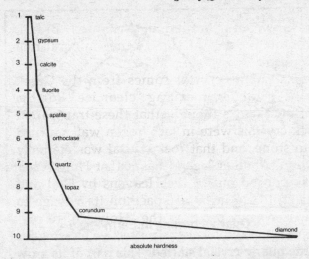

1 — talc
2 — gypsum
3 — calcite
4 — fluorite
5 — apatite
6 — orthoclase
7 — quartz
8 — topaz
9 — corundum
10 — diamond

absolute hardness

Mohs Scale of Hardness for the Ten Standard Minerals

atoms gives quartz its structured appearance. It occurs in an amazing variety of forms. In the world of gemstones, quartz supplies more varieties than any other mineral.

Much of the formation of quartz crystal occurred about 500 million years ago, before the land masses had their positions and shapes as we know them today. Upheavals occurred when land masses moved. Under the surface, loose sand and mud (sediments) settled into solid rock from the effects of heat and pressure. Sand became sandstone, and the mud became shale. These land masses moved and collided. The rocks buckled in folds and slid over one another. Large fractures formed, and inside these fractures fissures opened up. This is where much of

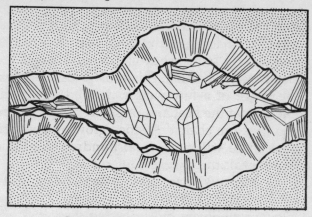

Quartz Forming in Earth Fissure

the quartz crystal started. This part of its growth probably took about 200 million years. Water seeped up into the fissures. Silica (from the sand) was in solution, and the quartz grew on the sandstone walls.

At 300 to 500 degrees Centigrade (which is the temperature necessary to form crystal), trapped water in the Earth which can't escape to expand and form steam turns water into a hot plasma of water and silicon dioxide. Very tiny quartz crystals which had previously been formed will then dissolve in this intense heat and pressure, much like sugar dissolves in a hot cup of coffee. What are left are crystal cells, and these cells have a memory. They "know" what they are and what they're supposed to do. They will swim around in this hot solution until they can find their "mother"—a matrix of crystal or an area where there was a small piece of crystal that had not dissolved. The cells next attach themselves to that piece and wait, and in a certain and orderly sequence would build a larger six-sided quartz crystal. This growth can occur in the space of a few weeks, depending on the intensity of the heat and pressure. Once the heat and pressure change, it stops growing and perhaps there will be several hundred thousand tiny crystals the size of a fingernail.

Scientists who are truly understanding of life have compared this crystal growth with that of the human soul. Every time a person dies, goes into spirit and comes back into another life, they leave some of their impurities behind and start over in a pure state. Both the crystal and the human soul grow into perfection by leaving unwanted impurities behind every time they recycle.

If a very large, perfect crystal is found in the ground, undoubtedly it is not a newly formed crystal, because they are formed over many thousands of years of growth (some may be only 500 thousand years old, while some are millions of years old), and each one generally has many impurities. They gather these impurities because the crystals start and stop growing thousands of times during their development. Each time the crystal starts growing, it traps the impurities on its surface and grows over them. (See the illustration of a natural crystal which has been sliced.) The cross section showing the lines look like tree rings, and gives a clue to the geological growth of that area.)

Geological History Recorded in Quartz Crystal
(Dark lines are the result of gamma radiation
when crystal growth was suspended.)

Location of Quartz Crystal

Most of the quartz found today in the United States comes from Arkansas. Although no other major deposits have been discovered yet, some experts feel there are many more areas of quartz crystal just waiting to be found, especially those areas where volcanic activity has occurred. In the Lake Superior area much amethyst has been mined. Others feel that Sedona, Arizona, has quartz crystal nearby. (Sedona is noted for being a vortex energy center and a gathering place for metaphysical people.)

In the 1800's, when the railroads were being built through California, small amounts of gold were often found where a tunnel was dug, Along with gold, large quartz crystal deposits were also unearthed, especially in Calaveras County. One crystal was discovered during this time that was reported to have been almost six feet tall.

Many other large, valuable crystals might have been found, but unfortunately the railroad workers would often blast away the unwanted crystals, as their purpose was to build a railroad, not to mine. The American railroad workers thought the Chinese workers were strange because they would pick up the pieces of crystal and ship them back to Chinatown in San Francisco. Chinese craftsmen would make crystal balls, and some would even be shipped back to China, where they were greatly valued.

Clues to where quartz can be found are given by other minerals. Silver and tungsten, as well as gold, can often be found in the same area where crystal is. Sometimes gold is actually found inside quartz crystal. When the interior of the Earth has been heated to 500 degrees Centigrade, the heat melts the fragments of gold which may come out as tiny gold filaments inside the quartz. Quite often when quartz ore is loaded with gold, it appears as a piece of gold lace or fern patterns that are spread throughout the crystal. It is very beautiful and expensive.

It is believed that there is a cave in which crystals grow located near Sedona, Arizona. Many psychics have felt that this is so. Lorraine Darr, who has proven herself to be an accurate psychic channel, confirms this:

"This is true. It is in conjunction with a transformation of the race, and it is in total cooperation with all of the sound flows that are in constant change, coming from what you call vortices, for these are implanted energies placed there by lighted beings from other spaces; placed there for use by the human race, and placed for the return of lighted beings to come into and enhance the human race when it is needed, when it is allowed, and when it is a moment of great change. There are at least three areas within 50 miles in which crystals grow in great numbers, and they could be termed 'crystal caves.'

"The crystal caves, if you wish to call them that, will attract those who will, as you might say, 'open to them' or 'find them' at points of need and direct change in this solar system. They're not just there for the finding and the taking. They are there as a focal point that participates in the growth and development of this area of the United States, and that also participates with many other well-known places around the Earth. One may be found in what you term 1993. But to spend your moments hunting for it is to lose the space that you have to find your own crystal cave in your heart....So, don't hide in your crystal cave. Open the door of it yourself. Know that you are crystal light in a form to rise upon this Earth in ways that you haven't yet thought of using, but you will."

Both quartz crystal and crystal caves have many teachings for us if we are willing to learn what gifts and lessons they have to share with us.

Growing Crystals

Growing crystals is enjoyable and educational. It is possible to grow simple types of some crystal forms. One easy experiment is to dissolve alum or salt in water until the water will absorb no more. Next, suspend a piece of string in the liquid, which acts as a wick. As the water evaporates from the wick, crystals will form around the string.

Quartz crystals will not grow in this manner, however, because they must have a type of system which can create controllable heat of up to 500 degrees Centigrade and 25 thousand pounds of pressure per square inch. It is extremely expensive to make equipment that can grow quartz crystals. The vessels in which they are grown in have to be contamination free. Scientists now use quartz crystal test tubes for radioactive carbon dating date because they are contamination free.

To grow high quality quartz, the chambers they are grown in are sometimes lined with pure platinum or silver so that absolutely no contamination occurs in the growth of the crystals.

There are several different crystal growing companies which supply today's electronic industry. Some of these companies are in Russia, Japan, and Germany. France also has one electronic laboratory which can grow crystals in different colors.

Colored crystals have value for metaphysical use. Most of their colored crystal is sold to the jewelry trade. They can make crystals in a citrine color, which if it is the right shade could be deceptively sold as topaz. The electronic quartz crystal could be said to be more valuable because of its effect on the human mind, although jewelers are less concerned with this aspect than are metaphysicians. The higher value should be placed on the activation of the human mind, however.

The laboratories today have perfectly reproduced Nature's method of growing crystals. It is exactly like refining scraps of gold into one large bar of pure 24 carat gold. Through heat and pressure the impurities are driven off, and Nature does the rest. No one can actually grow a synthetic quartz crystal. The only way growth can occur is by allowing Nature to recycle the crystal. What technicians do is present the best environment in which this can happen.

This method of growing crystals can be compared to growing a hothouse tomato. The tomato that grows wild in the vacant lot next door really isn't quite as luscious as it could be, because it doesn't get the right water, nutrients, heat, etc. Perhaps it was bruised and battered by dogs, birds, bicycles and playing children, but the hothouse tomato had exactly the right amount of nutrients, heat and water, so it grew into a large, healthy, perfect tomato. So it is with the laboratory grown quartz crystals. They have optimum growing conditions with consistent heat and pressure, and no impurities present.

These crystals are flawless and ideal for metaphysical or technological use.

Present Day Mining

As previously stated, quartz found in the United States today comes primarily from Arkansas, where it is found in red clay and is extracted from open pit mining. There are two grades or qualities: the common grade, and the specimen grade, which is clear. The white common grade (called "contaminated" because of the many mineral impurities in it) is used for commercial purposes. It was long considered to be useless, but now it is recycled for the electronic industry. The white quartz is first crushed, then soaked in oxalic acid heated to about 180 degrees Fahrenheit to be cleaned. The discarded pieces from this procedure are sold for landscaping material. The choice white quartz is used for feed crystal. These pieces are placed in the bottom of pressurized vessels. Into the top are placed thin slices of pure crystal called "seed crystal." Heat and chemicals are added, and in about one to three months clear, manmade, perfect crystals evolve. The heat and chemicals cause the white, milky crystal to dissolve, releasing their cells which then grow onto the seed crystal. Perfectly clear, uniform crystals are formed. These are cut into wafers for oscillators and other electronic applications used for TVs, radios, CBs, telephones, VCRs, satellite communications, etc.

Earthquakes and Crystals

Earthquake activity has a spectacular effect on quartz. Whenever earthquakes occur, they squeeze millions of tiny crystals in the Earth's crust. This squeezing generates a massive output of electrical energy which can result in a discharge of electrical lightning. Although there are millions of volts which go up into the ionosphere and dissipate, there is very little amperage created and so this lightning can usually only be seen at night.

When primitive people saw this, they became very terrified. They believed the gods were angry at them. Not only were they shaking the Earth in their wrath, they were sending sparks of fire into the already ominous night sky.

Every time a slight earthquake occurs, quartz crystal also sends out electrical signals into the universe saying, "Here we are." Even when the Earth is at rest, signals are being sent out from crystals in the crust.

Chapter Three

TYPES OF CRYSTALS

A crystal is a solid material composed of silicon dioxide with a regular internal arrangement of atoms. Because of this orderly composition, it forms the smooth external surfaces called faces which allow us to see into the crystal when it is clear. Quartz crystal always has the symmetry of the hexagonal system (in metaphysics it is often referred to as the trigonal system).

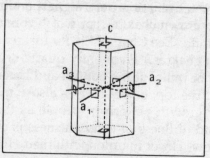

Hexagonal Crystal System

This system has one six-fold axis of symmetry, while the trigonal crystal system has one three-fold axis of symmetry. Both systems have the same set of crystallographic axes.

Most stones are made in part of silica. The presence of this silica is what gives crystals their luminosity and crystal clearness. Crystal is brittle—as we are—and as such is a reflection of ourselves. As we "shatter" our being (or come to really understand ourselves), it is seen to be rigid and crystalline in structure (our skeletal structure is rigid and unbending, and often we have a fear of change or are unwilling to change our mental process).

Crystals are described as being clear, milky, having rainbow prisms within, or having fractures visible along their length. When people talk about crystals, they are usually referring to clear quartz, the most common form, which is often called the "grandfather" of the mineral kingdom.

Quartz crystals represent the sum total of evolution on the material plane. The six sides of the quartz crystal symbolize the six chakras, with the termination point corresponding to the seventh crown chakra, that which connects with the infinite. (See Chapter Six for further information on chakras.) Frequently quartz crystals are cloudy or milky at the bottom and become more clear at the top, or point. This also symbolizes a similar growth pattern in people as the cloudiness and dullness of consciousness is cleared as one grows closer in union with their Higher Self. Within clear quartz crystals are usually found inclusions, or clouded areas. Sometimes these resemble galaxies and in a sense they are, for "As above, so below."

Right and Left Handed Crystals

Quartz crystals grow either right or left handed. The difference is caused by the positions of the spiral structures on their faces. Right or left handedness can be visually determined by observing the position of the small faces adjacent to the large pyramid faces near the point or apex. There will be a small face at the bottom of the large six faces and it will point or slant to the right or the left. In many cases the small slanted face will be missing, however it is still possible to classify a crystal as being right or left handed by placing the crystal in an alcohol bath. Under polarized light, it will increase in intensity if it is

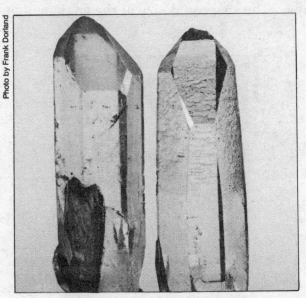

Photo by Frank Dorland

Right and Left Handed Crystals

right handed, and will decrease in intensity if it is left handed. Whether a crystal is right or left handed has nothing to do with its abilities or capabilities. One will work as well as the other for any purposes, *as long as it is clear.* This right or left handedness is just a technical consideration in describing crystals.

Single Terminated Quartz

The most common type of clear quartz crystal is the single terminated, six-sided column-like length of mineral. Six natural facets join sharply together to form the terminated apex. A crystal's angles are not always consistent at this apex. The shape and size of the angle will be

Photo by Frank Dorland

Single Terminated Crystal

dependent on the amount of space the crystal had when it was forming in the Earth. For instance, if it had to grow through a crack in granite, its tip could be somewhat lopsided. The most prominent faces will be those which had the most space for growth. Under ideal growing conditions, all faces will be equal.

Double Terminated Quartz

A single terminated crystal is formed when the six faces of quartz join together to form a point on one end. When both ends of a crystal join in this manner, a double terminated crystal is formed. Some crystal workers say that these double terminated points are useful in that they

Photo by Frank Dorland

Double Terminated Crystal

have the capacity to draw in energy as well as radiate energy from either end of the crystal. By uniting the energies together in the central body of the crystal, a double terminated crystal can then project that unified essence out from both ends. Instead of growing out of a hard rock surface where single terminations are formed, they grow in the center of softer clay. They know no limits or boundaries, and have grown to completion on each end. The double terminated crystals can teach that it is possible to be balanced in the dual expression of spirit and matter.

Quartz Cluster

A quartz cluster is a natural conglomerate formation of two or more single terminated quartz crystals joined in a massive rock matrix. A quartz crystal cluster can have as many as 100 individual quartz points or terminations, or they may have only two. These clusters share a common base. They are many individual crystals who are thought to all live together in harmony and peace. They represent the evolved community, each member being individually perfect and unique, yet sharing a common ground and common truth with the others. In clusters, as in every society, all units join together to reap the benefits of living, learning, and sharing. The individual crystals reflect light back and forth to one another, and all bathe in the combined radiance of the whole as we humans too must learn to live together in harmony on Earth.

Photo by Frank Dorland

Quartz Crystal Cluster

Tabular Crystal

A tabular crystal is somewhat flat in appearance. This is because two of its six vertical column-like sides are each at least twice as wide as any of the other four. This crystal may be either single or double terminated. The wide flat faces can be used as a sending and receiving board for telepathic healing and communication. Many psychics believe that because of the two larger flat sides, tabular crystals have a different energy frequency from any other quartz configuration, and are excellent to use for balancing any two elements, chakras, or people.

Photo by Frank Dorland

Tabular Crystal

Herkimer Diamond

Yet another unique form of quartz crystal is the Herkimer Diamond. A higher vibration or octave form of crystal, Herkimers are generally smaller, but of a nicer quality (clearer and more brilliant) than other double terminated crystals. They are different because when they were forming in the Earth's plasma, they had no place to attach themselves. They were formed in soft clay of their own volition with no restrictive surfaces to grow against. Thus they were free to grow closer to perfection. They were first discovered in Herkimer, New York. Recently a large deposit of this type of quartz crystals were found in Mexico. They are brilliant, usually double termi-

Photo by Frank Dorland

Herkimer Diamond

nated, many with small black flecks of Androxylite, many with rainbows.

Although they are small, they are very powerful. Herkimer Diamonds are considered to be catalysts of the quartz family. They have the potential to initiate deep inner transformation of the self.

Lead Crystal

Many ornaments of beauty made from lead crystal such as vases, glasses, candle holders, paperweights, sun catchers, etc. can be found in homes Although these objects have much aesthetic beauty, their attributes are not to be compared with quartz crystal. Lead crystal is not quartz crystal, but glass compounded with a lead oxide.

Lead is a radiation barrier, which is why it is used as a shield for x-rays and other forms of harmful radiation. Glass, an insulator, does not allow any electrical energy to pass through it, so do not mistake it for quartz crystal. Lead crystal will not give you any electrical amplification or feedback from your body cells; *only quartz crystal has this ability*. Enjoy lead crystal objects for their beauty and fine craftsmanship. They are also far less expensive than the same items if made from quartz crystal.

Colored Quartz

There are several other varieties of quartz which occur naturally. Colored quartz can be found in several interesting and beautiful hues. When chemical impurities are present, a crystal will be tinted in shades of yellow, rose, blue and green, but rarely are these of a high enough quality to be used as optical, electronic quartz crystal (these are usually translucent, but not transparent).

Although it is difficult to pinpoint exactly what causes the colors in transparent quartz crystal, scientists have suggested that certain minerals are responsible. Citrine may be yellow or brownish because of inclusions of colloidal iron hydrates. Rose quartz is a pink color thought to be caused by traces of manganese or titanium.

If crystals are exposed to radiation (such as the natural radiation which occurs in many places in the Earth's crust), an electron is gradually displaced in each atom of the crystal. This radiation colors the quartz, producing amethyst, citrine, or smoky quartz. Amethyst appears as various shades of a light to deep violet or purple color. Some mineralogists believe that the violet color of amethyst is caused by traces of ferric iron. Smoky quartz ranges from yellowish brown (or a light smoky color) to a dark, brownish black, sometimes a deep, mysterious, transparent black color. Citrines range from pale straw or champagne to lemon yellow, orange and red/orange (topaz color).

Electronic Quartz

Electronic quartz crystal is high quality (H.Q.) single point crystals which may be clear or colored rock crystal, but it is always fully transparent and flawless. Thus, the electrical impulses will always be consistent.

The perfect clarity of electronic quartz is due to the absence of any measurable amount of chemical contamination. When this type of quartz is squeezed and released, it produces tiny amounts of electricity, which oscillates or vibrates at a precise frequency. This controlled repeating vibration has made many technological achievements possible for our communications systems.

Presently, electronic engineers are not using H.Q. amethyst, smoky, citrine or blue crystals as well as the clear. However, there are reports that scientists working with solar panels are experimenting with H.Q. blue crystals for solar electrification and have had good results.

When an electrical charge is applied to quartz it oscillates. Oscillators (integral devices used in modern communications equipment) made from quartz regulate electrical energy. The frequency with which they oscillate depends on the size and thickness of the quartz.

Colored quartz is beautiful to look at, and many healers prefer colored to clear quartz for their use; but undoubtedly quartz crystals' best qualities are realized when they are electronic (perfectly clear). If all the colored quartz were heated above 300 degrees Centigrade it would come out colorless, and become clear quartz. Next, if these now clear quartz crystals were bombarded with gamma rays, either with an x-ray machine or a linear accelerator, they would turn back either to smoky or amethyst, etc., depending on their original color.

Amethyst can only be recycled from real amethyst. Laboratory-grown amethyst grows into a crystal brick which is colorless. Next it is bombarded with gamma rays. Whatever impurity in the structure that responded to gamma rays is still there, because it too accompanies the crystal cell that turns it back into an amethyst.

The most common form of rose quartz is opaque. This type of quartz is a beautiful rose color, but you can't see through it because it's full of iron oxide, which grounds it out. Translucent (allows light to pass through but is not clear) rose quartz is not electronic either, but it is still valued for its aesthetic beauty. This type of rose quartz is inexpensive and is excellent for such things as jewelry, a decorative crystal ball or a lamp base. Although it won't transmit any energy, it's beautiful to behold. Electronic rose quartz does exist, although it is rare. (Remember, for quartz to be electronic, it must be completely transparent.)

Translucent crystals work better than opaque ones, however. The opaque, cloudy ones are full of impurities and chemicals which interfere with their electronic transmission. Crystals with small cracks and fissures (inclusions) can still transmit energy. If you hold the crystal up to the light and you can't see through it, this indicates that it is so full of other minerals or chemicals that it is unable to transmit vibrations. In naturally grown crystals, the point, or apex, is the important part, and if that at least is clear then the crystal will be more effective for healing, metaphysical or creative use.

The most important aspect of quartz crystal at our present level of understanding is its receiving, amplification and broadcasting ability for the human body, mind, soul, and brain.

Any transparent quartz would respond to programming. (Programming will be explained in Chapter Five.) It is best to have one single crystal point, not a twin, nor one where several points grow together. One crystal point has one wavelength of vibration. A twin or cluster would have several different vibrations, just as two or three radio stations tuned in at the same time will interfere and cancel each other out. Clusters are beautiful to look at, but if you want a particular crystal to respond to your programming, you should take that individual out of the cluster, remove its points and all sharp edges, tumble and polish it, and use it alone. A cluster is best used for aesthetic appreciation in the five-sense physical world. It is beautiful and enjoyable, but it is not an effective tool because it lacks a single, focused energy source. Learning to focus and harness this remarkable energy source by polishing and smoothing quartz is the next step in our understanding and use of crystals.

Crystal Veils

As we know by now, the clear, transparent crystals are the electronic ones. The electrons are distributed throughout the crystals whether they're clear, citrine, smoky, or amethyst (the colored quartzes which are perfectly clear—like glass). Remember, the ones that have the impurities of chemicals in them are *translucent*, not transparent. Those very seldom have any electrical qualities because the impurities tend to ground out any electrical energy and they are ineffective as a working tool. Trying to use an opaque or translucent crystal for transmitting energy would be similar to throwing water into a good electrical receiver—everything would short out.

Some crystals have veils in them—small wispy clouds. The questions has arisen as to whether these have any useful value. Some of these crystals have dozens of phantom crystals in them too—pyramid shaped inclusions showing where they stopped and started growing several times.

Certain psychics prefer crystals with veils because they have read in older divination books that when the veils lift (or seem to disappear), the visions are seen. When you gaze at a crystal, the crystal will often turn cloudy and then will become clearer and clearer. Finally, you're not aware of the crystal at all because your five senses are absolutely subdued. You are not actually looking at the crystal with your two physical eyes; you're in a trance state and you're looking at whatever is psychically occurring.

Some people are fascinated with the different veils and striations in a crystal, and they prefer having this type of crystal. A veil usually appears when the crystal has stopped growing for two or three thousand years. It looks like "the dust of the ages." There would be a small amount of dust or impurities that would stick to the surface, and then when the crystal started to grow again, these of course would be included inside the crystal. Sometimes there might even be a little hollow spot with a drop of water (also an inclusion). As long as a veil or impurity goes through like a cloud, it doesn't reduce the crystal's effectiveness as a working or divination tool.

There is an added advantage to using these kinds of crystals for those people who hold the belief that the veils and small inclusions that sparkle in the sunlight are the souls of one's departed ancestors. Throughout much of the world, there is an ancient belief which states that one's guardian angels are actually their ancestors. Both ancestors and guardian angels have been considered to be protective helpers to those living in the physical world. Many ancient people would visit the graves of their ancestors and talk to them, hoping for advice on what to do. They relied on a departed mother or father, grandparent or spouse to speak to them through intuitions, hunches or dreams.

Again, some people prefer only perfectly clear crystals. Monetarily, there is a big difference. One that has veils is usually less expensive, except in some cases where in order to get a particularly good veil you have to pay more for that crystal, because it has an extraordinary, beautiful veil!

If you are making your own crystal working tools it is better to cut away the clouds or veils unless it is a particularly beautiful piece of crystal. The famous Mitchell-Hedges crystal skull has veils in the forehead which appear in a slant. Biocrystallographer Frank Dorland, who studied the skull for six years, believes that the skull has these veils simply because to cut them away would have greatly reduced the size of the raw crystal from which it was made. (For more information on this famous crystal skull, see Chapter Four.)

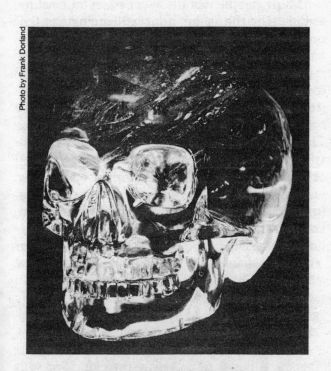

Photo by Frank Dorland

Crystal Skull Showing Veils in Forehead

Tourmaline and Rutile Quartz

Many people wonder what effect tourmaline and rutiles have on quartz. Tourmalinated quartz is a clear form of quartz with filaments and threads of dark tourmaline running through it. Rutilated quartz is clear quartz with threads of titanium dioxide (golden filaments) running through it, sometimes called angel's hair. The filaments look like gold or silver colored threads. These threads can also be rose gold or peach silver. The best specimens look like fine spun gold or silver baby hair. An engineer in electronics wouldn't use a crystal with tourmaline or rutile in it for an oscillator or any other electronic unit, because the electrical vibrations emitted would not be consistent, whereas high quality quartz

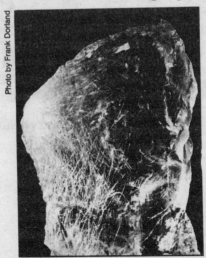

Photo by Frank Dorland

Quartz Crystal with Rutiles

(HQ) would be consistent. This type of clear, electronic quartz crystal is grown in laboratories under very exacting conditions. For metaphysical uses, crystals with titanium and rutiles work as well as clear quartz. Psychologically, some people prefer these inclusions because of the aesthetic beauty. Some healers even believe these crystals work even better, perhaps because of their preference for this type of healing piece.

Polished Crystals

The single terminated pointed crystal that comes out of the ground is very sparkly and beautiful. Mother Nature makes a crystal in this form with its natural terminations, and this is her method of communication and energizing herself. This type of crystal puts out electrical impulses. Although no one knows for sure, these impulses that are emitted from quartz could have a great effect on many things in our world. The vibrations probably go throughout the Earth in the area where the quartz is, and it could trigger certain responses, growths, etc. in other crystals, in plant life or anything else that is in contact with the ground.

The energy that comes out of these crystals is very definite, but it is also very slight. Currently there are no instruments sensitive enough to measure and understand what is coming out of the crystals, but we do know it is being emitted.

Now Nature has a way of protecting and sustaining herself. When these crystals are buried in the Earth, the Earth surrounds them and with the ever-present moisture in the ground makes contact with the flat surfaces, so the energy which is emitted can flow right back into the ground. When the crystals are mined and removed from the Earth and washed and cleaned, they reflect light and color, and have a very beautiful appearance—but most of the energy is locked inside. Crystals function in different ways when they sit in a room and you ask them to participate with you. Even if you don't communicate with them, they still do trigger things, and awareness occurs, but so much more happens when you speak to them, and even more interaction takes place when you pick up a crystal and hold it in your hand.

A naturally grown, naturally "faceted" crystal works most effectively on the five senses because it's broadcasting or relaying a mirror-like image of this five-sense world. This type of crystal can be likened to a hall of mirrors where you're on the outside seeing all of these different images (broadcasts). The energy that is built up inside the crystal is mostly wasted because it is inside its "hall of mirrors," similar to multi-wave television broadcast. The crystal vibration bounces around the six sides and wastes most of its energies because it can't efficiently get out. There is a method, however, which will effectively use most of the energy from this type of crystal.

Hand Working Polished Quartz Crystal

Photo by Frank Dorland

If you have a naturally grown crystal with flat sides and a sharp point and you want to effectively utilize its energies, wrap your fingers around it each time you use it, because that breaks up the flat surface so that it no longer functions like a mirror. When you completely surround the column of crystal with your hands and fingers, it actually becomes a reflection of the body cells and will make it easier for the energy to flow in and out. Used in this manner, the crystal's energies are much more accessible and efficient.

There is another speculative reason why crystals may grow as they do—faceted and pointed. Maybe Nature had a way of protecting herself by having this potential energy hidden. Anyone who buys a crystal point (natural single or double terminated) and tries to do black magic or use the crystal for unethical purposes would have limited success, because the crystal would have very little power. The rounded and smoothed crystal represents a higher level of working. The rounded healing tools are also more expensive, because they require much more time and labor to fashion them.

Crystals are here for our use in learning more about ourselves and our capabilities. Perhaps the sharp crystal point or end is created in this form to attract people's attention because it is so beautiful.

Crystal Magic

The following story is an example of how natural pointed crystals could have been used to impress early people:

If a shaman or medicine man fastened a pointed crystal on the end of a stick, much like a fairy godmother's wand, and then twisted and turned it, this would look very impressive.

There is a cave in the western United States used by early Native Americans, that could only be accessed through a long carved-out tunnel going down into the ground. When the sun was at a certain point, a narrow band of sunlight would shine through and light up a precise area, where a hollowed-out depression was located.

The shaman would conduct ceremonies in this cave. As part of the ceremony, the members of the tribe would eat hallucinatory mescal buttons. This would give them Divine insights, but would also contribute to the psychological control and impressiveness of the shaman. When all the tribe members were present and in an altered state of mind, he would use his stick with the crystal fastened on the end of it. When he placed the stick into the depression and twirled the wand around, the sunlight would be reflected on the crystal and the light rays would bounce all around the walls of the cave. Flashing this and foretelling future events, he maintained his power image. If he was a good shaman, he would use these theatrics to help his tribe, because the people would believe anything he told them. The tribesmen would think that the shaman was actually harnessing the power of the Sun.

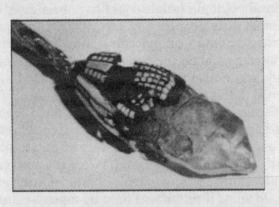

Ancient Quartz Crystal Witch Doctor's Wand
(from museum in Santa Barbara, CA)

Crystal Vibration

All living things have a certain vibration or frequency. A psychically sensitive person can pick up a rock or a piece of jewelry, for example, and tell where it has been, the experiences it has encountered, and can discern a feeling from it. Anyone who is sensitive and wants to experiment with their intuitive abilities can hold a piece of jewelry and get a "feel" from it, or impressions about its owner.

A crystal has its own unique vibration, also. Lying on a table or placed by itself, a piece of crystal has its own special frequency. The Chinese call quartz the Living Stone, because they can sense a pulse in it. When you pick up a crystal and hold it in your hand, the vibrations change. It is no longer vibrating at its own frequency; it automatically begins vibrating on a harmonic with your energy source. Because you are feeding energy into the crystal, it automatically responds to you.

It is as if you suddenly have a choir and added to it the bass, alto and soprano. They are all joining you in harmony at a regular rhythm. The harmony and the amplification that the crystals bring about boosts your intuition, hunches and perceptions. All are being heightened throughout your body, so instead of suspecting something, you *know* it.

If someone else picks up the crystal after you have been holding it, it immediately changes to their vibrational frequency. Some crystal workers maintain that there's a certain crystal meant for you that will vibrate with you, but each crystal will vibrate with whomever is holding it. Some crystals may be too powerful for a given individual at a certain time. If you do not program a crystal to benefit you, it may just magnify the feelings that are occurring in you at the time. If there are a great deal of negative emotions, the crystal will magnify this.

Sometimes it is excellent psychology and salesmanship to tell someone that a particular crystal is made exactly for them! Some crystals will automatically respond faster than others because they happen to be the right density, weight, etc., so when you pick up that crystal it can easily tune in to your body cells faster than another crystal which might be larger, more sluggish, and have a lower natural vibration.

Some crystals react faster due to their individual natural vibrations. Ten different crystals could conceivably have ten different frequencies, depending on the size, thickness, concentration of the contents, and the impurities within each.

Quartz crystal is the basis for our present electronic communications system. Without it, the development of our communication would be severely limited. Everything is dependent on the fact that you can control one particular vibration and extract it from all the millions of vibrations in the universe, and utilize that vibration for communication from one person to another. This is possible through the use of quartz crystal.

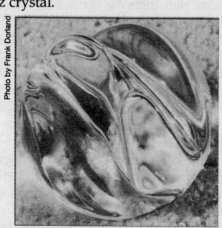

Photo by Frank Dorland

Large Hand Carved Quartz Crystal

Chapter Four

CRYSTAL BALLS

A book on crystals would be incomplete if it didn't contain information on the most famous form of crystal, the crystal ball. This unique form of quartz has been found by archaeologists in such widely separated areas as Peru, Siberia, Australia, Chaldea, Greece, Rome, Assyria, Persia, Japan, China, France, Ireland and Scotland.

In South America, Yucatan diviners looked into a clear stone they called zaztum for their visions, while on the other side of the world Tibetan monks called crystal balls the "windows of the gods," using them as holy objects of great power. The Taoists believed that looking into the crystal's clarity "crystallized" one's being, and considered quartz the "gem of enlightenment." Buddhist altars included quartz spheres as an invocation of the "visible nothingness" that delineates the duality of the material and spiritual world.

Contemplation of this "visible nothingness" gave rise to crystal gazing, which has been practiced since time immemorial—no one really knows how long—probably at least 15 thousand years. Crystal gazers use the spheres as windows to faraway places, and to the past and future.

Some of the early scrying spheres were not even made of quartz, with which we commonly associate crystal balls, but were fashioned from beryl, a member of the crystal family. It has a hardness of 7.5 on the Mohs scale, and is also found throughout the world. Transparent specimens are rare, however. Other scrying balls were made from aquamarine and emerald, also members of the beryl family.

One of the first optically round crystal balls was used by the famous Dr. John Dee, court astrologer to Queen Elizabeth. It is said that he used his crystal ball to choose the best day for the coronation of the queen. He spent long hours peering into his crystal in order to get archangels to advise him on certain matters. Dr. Dee's crystal ball was made from smoky quartz and can still be seen today in the British Museum in London. Previously, older scrying spheres were somewhat lopsided and imperfectly round.

Many other small crystal balls have been found in England. They were used for good luck and healing charms. Most of them are very tiny, some thumb-sized. They were highly valued and passed on in families from one generation to the next. (One of the world's largest crystal balls weighs over 100 pounds and can be seen in the Smithsonian in Washington, D.C.)

Due to the common belief that quartz crystal was frozen ice, fashionable women in ancient Rome carried quartz crystal balls in their hands during the summer months to help keep them cool. In past times, when the high priest of the Buddhist or Hindu temples would conduct special ceremonies, he would wear a consecrated robe and one or more small crystals of dazzling brilliance suspended from his neck on a beautiful chain. The priests believed that the crystal enhanced their ability to communicate with the spirits who helped them with solutions for their problems.

In some cultures, young boys and girls were used for their abilities of divination, as no one else was thought to be pure enough. The seer would intone prayers over the crystal before handing it to a youth or virgin. It was thought that they were able to read the answers to questions asked, sometimes conveyed by means of written characters on the crystal, and sometimes by the presence of spirits or angels who appeared within the crystal's depths (Dr. Dee used this method).

In other cultures, the ceremony which preceded crystal gazing was elaborate and spectacular. It included such accoutrements as a holy table or Lamen, swords, wax candles in highly polished gilt or brass candlesticks, and many other objects associated with magical functions. Frequent washing and cleansing preceded these rituals, along with prayers uttered by priests or adepts trained in crystal reading.

In many of these situations, crystal scrying was done when the Moon was waxing (on the increase). Some people felt that the greater the increase of the Moon, the greater the supply of the Moon's magnetism in the crystal. In astrology, crystals are said by some to be influenced by the Moon, as are the intuitive powers of the mind.

Crystal gazing was also employed by the Egyptians, who also believed that the crystal had magical powers. Often they would place a small piece of rock crystal in the center of the forehead of a corpse before mummification. This crystal placed on the third eye area would then serve to guide the deceased throughout eternity.

Dragons and Crystal Balls

It is perhaps in the Orient that crystal balls are most depicted in art. There are numerous instances where dragons are shown clutching crystal balls in their mouths or with their claws.

The ancient Chinese and Japanese regarded quartz as the perfect gem. The artists who carved spheres were thought to be the most capable of spiritual and artistic purity. They considered the quartz crystal ball to be the heart or "essence" of the dragon, symbolic of the highest powers of creation. The Chinese and Japanese shared the term *sui ching* for quartz, which means

Photo by Frank Dorland

Dragon Holding Crystal Ball

"water essence," the source of peace and power. Japanese lore originally proclaimed that rock crystals were the gift of magical dragons. Small crystals were believed to be the congealed breath of the great White Dragon. Larger, more brilliant crystals were said to be the solidified saliva of the dramatic Violet Dragon.

Many dragons in art are portrayed with the crystal in the mouth. The crystal ball represents the dragon's breath—pure, undefiled absolute Truth. There is another legend of a dragon king who lives in the sea in his crystal palace. He has an army of crabs, shrimp, etc. who patrol the ocean bottom for him.

Ancient History of Crystal Ball Usage

Crystal balls are thought to be first used from 15 to 18 thousand years ago. The following scenario is an example of how they could have been initially discovered:

A person of a very ancient culture, wandering through a riverbed after the spring runoff would be out looking for food, scavenging. Of course he would walk through the riverbed because by now the water had dried up, and it was much easier going than climbing around the cliffs. Every once in a while he would find a piece of quartz crystal that had fallen down from the cliff. The waters rushing over it and the sand in the riverbed had rotated the crystal and tumbled it so that all of the points were worn off and

Photo by Frank Dorland

River Tumbled Crystal　　**Lopsided Crystal Ball**

Photo by Frank Dorland

**Three stages of making
a crystal ball**

formed a lemon-sized cobblestone piece of crystal. He would also find crystals that hadn't fallen into the riverbed which were even more beautiful because the sharp, clean sides and angles would reflect the sunshine and flashes of light.

However, he found out that when he picked up the smoothed crystals which he had found in the riverbed, he suddenly started receiving images, mental impressions of things; it amazed him to have a direct communion with the Divine. He realized that he actually had a discernible reaction! His intuition was heightened, his perceptiveness was crystallized into a vision that was very vivid, and the clarity of his mind was now very sharp. He therefore reasoned that it was those crystals which were rounded that worked the best. This was a round stone like no other stone! The sunlight would go through it.

The Sun was part of God, because when the Sun went down it seemed like God had left them. When it came up in the morning, often some of them weren't alive any longer—the darkness had taken away their life overnight. (Many ancient cultures equated the Sun with life, warmth and growth, and the darkness of night with coldness and death.) The crystal would be regarded as a divine object. They would hold it in their hand, and they suddenly understood things that they hadn't realized they knew. This was the start of the crystal ball used as a divining tool.

The use of crystal has probably been universal, except that it has been restricted to the secret brotherhoods, royalty, and select people. These individuals found out that their intuition was heightened and that they survived for a longer period if they had a crystal. Quartz crystal was really a survival tool.

Many cultures' royalty have used crystal balls for thousands of years; no one knows for sure how long. The French in particular are noted for their use of crystal balls. A perfectly round crystal ball was found in Childeric I's grave.

He was one of the first of the Merovingians, and the father of Clovis I. Childeric lived c. 437-481 A.D. He worked with crystals and magic. His kingdom was one of the most beneficial kingdoms ever in the history of France, and it was successful in repelling unwanted invaders, like the Visigoths and Anglo-Saxon pirates, from the kingdom.

How to Use a Crystal Ball

Use of a crystal ball for gaining intuitive information is so popular because it is the easiest method known by which astral vision may be trained. Quartz is the best to use because of certain atomic and molecular arrangements which tend to promote psychic powers and faculties.

Many people are natural clairvoyants, and after a short time of practice with a crystal ball readily see with the mind's eye impressions received in the form of a picture from within the crystal. It is easier than supposed to concentrate on the thought which the visions or forms conjured up.

Although the traditional perfectly round crystal ball comes to mind for divination (scrying) purposes, you can use any piece of crystal for scrying. Of course a rounded, smoothed crystal will give you better results. The size is less important than the clarity of the crystal, and even a small crystal of one or two inches will serve to focus your attention.

It is best to be in a relaxed, meditative state (the alpha or theta level of brain wave activity) before attempting to gain information or seek answers from a crystal ball.

Pick up the crystal ball and hold it in your hand to "turn it on." Look at the round surface,

Photo by Frank Dorland

Traditional Crystal Ball

and then project your mind *through* the surface and into the crystal. Put your mind inside the crystal. What happens then is that you are shutting off your five senses and tuning into your sixth sense, or intuition. These five senses are being subdued down to the point where you have a channel going into the crystal (sometimes called an astral tube or a line of force).

Now you are feeling the crystal's vibration and this is amplifying reactions in your body cells, which are a part of the communications system. Whatever you wanted to know comes up to your conscious mind in the form of visions because your mind is in the center of the crystal.

You are actually trying to project your mind into the crystal. A rounded crystal sphere is just as effective as an optically perfect ball. Some are like an orange or grapefruit—they aren't perfectly round.

You have probably seen pictures of the old traditional circus fortunetellers for hire. They are shown with a great, enormous crystal ball which is very impressive, atop a fancy stand and their hands are on either side of the ball. (The crystal ball is much more effective if you actually contact the crystal by touching it with your hands.) The fortuneteller doesn't really want to touch the crystal because then she will have to take the time to wipe off her fingerprints. She wants the crystal to look perfect for her next customer, who is waiting outside the tent with ten dollars.

When you contact the crystal, fingerprints are unimportant because you are looking *inside* the crystal. It is correct when you say "What do you see inside the crystal ball," not *on* the crystal. The surface represents the five physical senses (reflections). You want to go inside where you have the communication with universal intelligence, with all aspects of your selves: your Higher Self, your subconscious, Superconscious and even the metaconsciousness.

Gaze at the crystal and put your concentration in it. Quite often the crystal will look cloudy. It appears somewhat like a vision that is blurred. Next, the vision becomes clear and the crystal itself will seem to disappear. Most everything will disappear, because at that time you are in a daydream or trance state (alpha state) where you see visions and hear voices, but not with your two physical eyes. Experiments have been done with a crystal ball in which someone was seeing images and the crystal ball was photographed. The camera didn't see anything because the camera doesn't have a brain! That's the difference. Whatever is seen in a crystal ball isn't necessarily true. It depends on your personal interpretation of what's going on, and the clarity of your own channel, your own consciousness whether you are confused or whether you are balanced, grounded, integrated at the time.

A polished surface is important because it enables you to go past the surface into the inside of the crystal. (Even one with a flat surface will work.) A surface full of scratches will make it difficult for the mind to project itself inside when your physical eyes are distracted by the scratches. Remember, a small, polished crystal of an inch in diameter is suitable for scrying. It is just as good as a very expensive crystal ball.

Use of the Crystal Skull for Scrying

The Mitchell-Hedges crystal skull itself is one of the oldest scrying instruments in the world because the pate or dome of it does not have the suture marks of regular cranial bones, and the rounded dome is fairly smooth. All the other details are there, very precisely carved, but it doesn't have any anatomical details on the top. The Y that appears in everybody's skull would have been the easiest thing to carve into the skull. Because this area of the dome is so smooth, it is reasoned that it was used as a divining device by the priests, who would gaze at the forehead and enter a meditative state, much like looking into a crystal ball.*

Another reason why the skull is thought to have been used for divination is because there's a secret hidden reading glass carved into the upper roof of the mouth. A priest looking down into the skull from above could read the magnified written messages to his people. If the priest had an accomplice underneath the skull, hidden from others inside a hollow altar, this helper could write a message which would be seen only by the priest standing above looking down into it. There

---

*For more information see *The Message of the Crystal Skull* from Llewellyn Publications.

could have been a conspiracy of the priesthood to write certain things for the head priest to say, or perhaps they wanted to choose someone out of the audience for various reasons, e.g., to make an example of someone for a reward or to be sacrificed, thereby ensuring that the priest's power would be maintained by fear and awe.

The crystal skull wasn't originally a device to con the unsuspecting people, but in its later use when the priesthood degenerated, it undoubtedly was used for this. The priesthood realized they had something that was a powerful instrument, and that they could gain power, wealth and prestige with it. They discovered that they could enhance their own standing if they put God's voice in the crystal skull and put words in God's mouth, so that's what they did with the contrived messages.

To further impress the people into thinking that the skull was talking, the lower jaw was cut away from the rest of the skull. With a puppet-like push rod made of crystal tied to the lower jaw, the priests could then move the jaw up and down so that it appeared the skull was actually talking.

The eye sockets are also fascinating focal points. Because of the convex and concave structure of the carved in the eye socket, energy can be drawn in and sent out. Researchers say that the eye socket produces more energy than any other part of the skull. Because it is so mysterious in appearance, these eye sockets were likely used for divination.

Photo by Frank Dorland

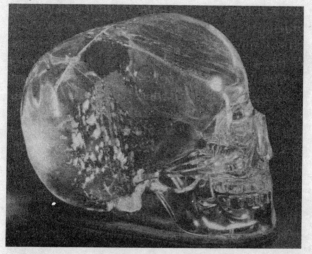

Mitchell-Hedges Crystal Skull
(showing smooth cranium with no carvings)

Perfectly Round Crystal Ball

Modern Crystal Ball carved for finger grooves

Chapter Five

PROGRAMMING CRYSTALS

Choosing A Crystal

How do you choose a crystal that's right for you? How can you know which one will help you? The best way to determine this is to listen to your inner self (intuition or gut reaction) and select the crystal that's interesting and attractive to you.

It is helpful to handle and hold a crystal, being sensitive to how it feels. Pick a stone by attraction, by being drawn to touch it, and then hold it in the left hand loosely. Let it rest against the palm chakra and notice the impressions, colors, sounds, and mood feelings that accompany the stone. No two crystals are the same. They are like snowflakes. Each crystal has its own particular vibration, and each will resonate differently. Look for one that will resonate with your being —regardless of size, shape, color, or jeweler's quality. It's the intuitive impression which is important.

The crystal should feel alive in your hand, vibrate or radiate. In other words, it should feel good. Even though each crystal has its own unique vibration, when you pick one up it begins to vibrate in harmony with you. Some will harmonize more quickly than others to your own vibrations, and these will feel good to you.

A tiny quartz crystal no larger than a thumbnail can be powerful and effective. Look for pleasing colors, clarity, and unchipped points, as well as symmetrical shapes. It is said that crystals are like cats in that they select their owner! If you are supposed to have a certain crystal, likely it will come to you somehow.

Here is an interesting technique if you want to check out the energy of a crystal. Wait until you feel quiet and centered in your heart, then hold the crystal in your left hand. Curl your tongue so that its tip touches the middle-back of your mouth. This area joins the energy of your two major acupuncture medians, and increases your sensitivity. When you touch this portion of the mouth while holding a crystal you may feel the crystal activate an organ, a chakra (see chapter six for more information on chakras), or some part of the body. You may feel a shift of consciousness, or simply feel warmth or tingling in your hand. Give the crystal a few minutes to speak to you in its own way.

Cleansing Crystals

There are many viewpoints on crystal cleansing, and whether it's even necessary or not. Follow your own feelings/intuitions on this. If you do feel that your crystal needs cleansing, then it probably does. Here are some methods of cleansing and clearing crystals. Use whatever feels right for you.

Use spring water and mix salt in it—preferably sea salt. Use a teaspoon to a tablespoon in a cup or more of water. Let the crystal soak overnight. Some say that salt is too harsh and that it removes energy but is all right to use on a crystal that has much negativity attached to it. Others say that salt seals the crystal and traps the energy within it. (Cleansing a crystal has been compared to swimming in salt water—it leaves your skin coated, but swimming in fresh water leaves you feeling refreshed and exhilarated.)

Caution: You can run water over your crystal to cleanse it, but remember that it is dangerous to plunge a crystal into cold water immediately after working with it or meditating with the crystal.

A large, good crystal when used in meditation or healing can generate a great deal of energy, and it can become quite warm—not hot, but very warm all the way through it. If plunged under running cold water, this would discharge its energies very rapidly, as the water acts as an electrical "ground." The crystal is also subjected to a thermal shock, as crystal neither spreads nor dissipates heat or cold. What it can do (and this has happened to some healers) is crack in several pieces, and this can make its owner very unhappy—and angry.

It is best to let a crystal rest after having used it for any length of time, or after having used it with any intensity, so that it can discharge its energies overnight and not immediately as it would when placed under running water directly after its use.

Some people like to cleanse their crystal by just holding it in their hands and visualizing it being cleansed by light, or by imaging the water from the Niagara Falls flowing over it. Crystals love to be out in the sunlight, because they are radiant energy sources. This is an excellent way to restore their energy. You owe it to your stones to give them some Sun! Many crystal users like to clear their crystals by resting them in rose petals. It is thought that the rose essence strips only the negativity from the crystals.

If you have a piece which has a particularly strong charge that you want to eliminate, you can bury it in the moist ground. If you live in the desert, water the immediate area it's buried under. Give the crystal a week or two (minimum of three days) to ground out.

When you dip or soak your crystal in water, bless the water. This will transform its energies into usefulness and thanks before you pour the water on a plant or onto the Earth.

One of the finest and easiest methods of crystal cleansing is to wipe it clean with a soft cloth. Hold the crystal in the right hand. Rest the right hand with the crystal in the left hand, and hold the crystal in the bright sunlight. Talk to it.

Order it to be completely cleansed. Take command, and do not be "wishy-washy." The crystal will respond to focused thoughts better than to vague, scattered thoughts.

Overcoming Old Programming

In ancient days, people believed that all of the energy came from the spinal column. This concept was used by the ancient Egyptians to protect their tombs.

They would carve the lower five vertebrae of the spinal column in a piece of quartz crystal, and hide this crystal in the west wall of the tomb, the direction from where all evil was supposed to originate. The priests would conduct a long, involved ceremony whereby the crystals would be programmed with the message that if anyone would desecrate the tomb, he would be instantly cursed and die.

The ceremony and crystal programming could explain why some explorers and archaeologists who have gone into tombs have met with untimely deaths soon after—such as in the case of King Tut's tomb. The effectiveness of this programming would be dependent on the subconscious programming of the individual.

Suppose an English explorer gained entrance to an Egyptian tomb. He discovered among other artifacts a carved piece of crystal. He picked it up and held it in his hands and admired it, and then put it down. The crystal would now begin vibrating from the programming instilled into it thousands of years ago at that final ceremony before the tomb had been sealed. Now that the explorer had picked up the piece of crystal, it had an energy source. Its "switch" had been flicked from off to on, much like a radio, and it now started vibrating to its ancient but still active programming.

The vibrations would go into the body and then into the brain cells of the British explorer. If he had an ancestry that subconsciously understood ancient Egyptian, his subconscious would respond, "Oh my god, we're cursed! Listen to that! We have defiled the tomb." The body would listen to the vibrations (the electronic messages), decode them and understand them. When the message had penetrated into the body it would respond by saying, "This physical being has to die." Immediately the subconscious would work to bring this message to reality by throwing the body chemistry out of balance and making the person sick. Perhaps the blood pressure would go up, or the pulse might speed up and slow down uncontrollably. The digestion might stop, and sure enough, the person would succumb and die.

Now suppose he had no ancestry at all in this lineage, didn't understand anything about ancient Egyptians, and couldn't care less. All that would be transmitted through his body from the crystal would be a lot of static, and nothing would happen. The curse wouldn't bother him in the least. In other words, the curse could work, depending on circumstances, the subconscious memory bank and a person's background and receptivity.

It is important and helpful for you to understand your background and where you came from. Much of what you believe is controlled by your subconscious memory bank, plus that of your grandparent, great-grandparents, and great-great-grandparents. Many people have gone through life perpetuating the same mistakes, the same faulty negative thought patterns their grandparents lived by. This happens because few people understand how to break this unnecessary subconscious programming about themselves. They have just repeated the stimulation which was in their genealogical memory bank. It isn't necessary to have ever known these ancestors or heard their voices—the programming carries on from generation to generation unless something is done to stop these mistaken beliefs from perpetuating themselves.

The crystal is one of the best and one of the least expensive tools to help you overcome this faulty programming.

Reprogramming Your Subconscious

You are born with capabilities you have never dreamed of! You can overcome the negative aspects of your genealogical programming by getting in touch with your subconscious—by telling it that *you're* in control, and that no longer will you respond to its old, oftentimes defeatist programming automatically.

One of the best ways to contact the subconscious is with the pendulum. You can make a pendulum yourself, out of a carpet tack made of iron or steel. Copper tacks will not work because they cannot be magnetized. (Any nonmagnetic material or substance lacking positive and negative cells makes a poor pendulum). Magnetize it, and suspend it with a silk thread two inches from your fingertips over a target card. This card should have numbers and the alphabet on it so the subconscious can spell out dates and information, and give you positive and negative responses. The pendulum, controlled by your subconscious, will swing backward and forward.

The very best pendulum is a crystal which is crammed full of positive and negative cells that greatly help in responding to the positive and negative in every body cell.* The pendulum should be tiny, not weighing very much so it can best respond to all the pulses of the body cells. Silk thread has a damping effect—that is, it allows electronic energy to go down the silk thread, but only so much. It hits the crystal, and it also picks up energy from the aura that surrounds the body—the magnetic flux field.

It is necessary to get to know your subconscious and become friends with it if you're going to communicate with it and gain information from it. If you are serious about working with the

*See Llewellyn's forthcoming *Crystal Pendulum Divination Kit.*

crystal pendulum, it is helpful to have both a healthy skepticism and a rational outlook in order to deal with the unusual answers which will be received.

A fascinating aspect of the subconscious is that *it* has survived (so you can survive) for so many lifetimes using trickery and deceit. It has needed all of its sometimes devious survival skills in societies that were not always kind. These habits are hard for it to relinquish, but reprogramming it with positivity is a challenge and is fun.

It is helpful to realize that we each create our own reality. While we are here on Earth we create our own heaven or our own hell, depending on our belief structure and that old programming we brought along with us. We are either being kind to our body and helping it to work properly, or we are destroying it through wasteful living or harmful attitudes.

The crystal is immeasurably helpful in amplifying the positive reprogramming, and it will help us to realize actual fulfillment while we are physical beings. Then when we pass on into the spiritual realm, we will have progressed and been properly prepared so we won't have to come back into the physical body and make all the same mistakes over and over again, as undoubtedly the soul had been doing in previous lives.

Unfortunately, the subconscious understands negativity and thrives on it. For programming, always phrase your changes in positive terms. Use positive affirmations so your subconscious can't trip you up and stay on the defeatist path it has been accustomed to follow. A popular affirmation is, "Every day in every way I am getting better and better."

Number of Programs in a Crystal

Many people are curious as to how many programs can be in each crystal. It really doesn't make any difference. From an electronic viewpoint, one will not cancel out the other. On the other hand, it is not a good idea to program the same crystal with conflicting commands.

If you program a crystal that you will lose weight and become slim and trim, certainly you would not want to program the same crystal the next day saying that you wish to be curvaceous and plump. All the programs should be beneficial to yourself, and be of a similar focus. An analogy would be like having a cassette tape with soothing music and one jarring, hard rock song on it. The rock music would tend to undo the serenity of mood that the other songs had elicited.

Clearing Crystals

One of the easiest and the most sensible ways to clear a crystal is to hold it in your right hand and then cup the right hand in the left hand so there's good electrical contact. Next, hold your crystal in the sunlight and try to clear it. Do not attempt to do this through window glass, because the ultraviolet light (which is what you want) is blocked by the glass. Ultraviolet light has a kind of cleaning and flushing action that passes through a crystal, and the frequency of this light vibration is what best cleans and clears quartz crystal.

So, you hold it in your hands, talk to it and tell your crystal that it's clear of all previous information. A programmed quartz crystal is like a tape recorder—when you clear your crystal, it's the same as erasing a tape. There's no more data left in there. Next, you can program it with something new as soon as the old program has been flushed/cleared/erased by the sunlight and your audible command.

What you're actually doing has less to do with the crystal than with your own mind and your own receptor vibrators, because the crystal isn't a complex, thinking, reasoning being such as we are. The human body and the human nervous system are the most complex system in the world, and that is what you're actually programming. Since the crystal amplifies all of what you are thinking or visualizing, it is called programming the crystal, but in reality the crystal doesn't think or care as we know it.

An example of a programmed crystal's effectiveness in protecting you: you put on your crystal (which you've previously programmed that only good things will happen to you), and you take a walk down the street. Someone who is walking your way who doesn't like your looks may possibly say something like, "Who's that jerk?" You've already programmed the crystal that anything beneficial goes into you but any negativity around you just dissolves and evaporates, and is harmless to you. Therefore, the negative thoughts of this other individual never enter into your body. Instead they go directly into the crystal, right into your system up into the hypothalamus, the central switchboard of the nervous system, but then is flushed away, discarded, put into a bucket or hole in the ground, sent to the Sun to be dissolved—whatever your favorite mental disposal technique is.

The hypothalamus is located in the most protected spot in the whole body, right underneath and in the middle of the left and the right brain, almost in the center in back of the third eye. It's very well protected because nothing happens in your body unless it goes through the nervous system, and the nervous system goes to the hypothalamus which is the switchboard network that knows exactly what to do with every message, where to send it, and where to store it.

The crystal will of course receive any messages that come along, but when it's programmed to accept only positive thoughts, that is all the hypothalamus will accept. Almost anything that you tell your hypothalamus to do automatically happens, so that the harmful thought patterns don't enter into your body system for any length of time to be stored in the cells at all, nor are they filed away in your sub-conscious memory bank. It is extremely difficult to talk to the hypothalamus directly, but you can easily communicate with it through the crystal, an effective and sensible method of reprogram-ming your body for health, happiness and success. Some adepts are able to talk to every single organ in their body if they are advanced enough, but for most of us, utilizing the crystal's amplification abilities ensures communication to the correct source.

You can tell all the different organs what you want them to do, to be of benefit and to activate themselves and be more energetic, or you can

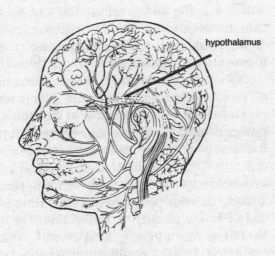

hypothalamus

Hypothalamus
Nexus of nerve fibers lining the pituitary gland.

talk to your hypothalamus. However, the best method is to tell your requests and needs to the crystal because the crystal amplifies and sends this electronic message. This message is transcribed as many tiny bleeps to your hypothalamus, to every cell of your body, and to your subconscious mind. All of these parts of your body receive the message at the same time, and they understand completely what all of those little bleeps mean. They have full understanding and the ability to decode those messages. The human body is truly amazing. Your human body is indeed a marvelous radio-like message receiver. The only thing you don't have is an amplifier. You have a very efficient antenna system which is the human body. You also have the tuner, which actually is the mind. You can select and tune in what you want. Your five senses (your sight, your sound, your hearing, etc.) can be likened to a television set.

What you don't have is the amplifier to boost all these messages that come in. This is what the crystal does: it amplifies and boosts up that power of understanding and brings it into your five-sense conscious awareness so that now you can actually be aware of ideas, health problems, unresolved situations, etc.—things that have already been in your body all this time but you didn't know about, or know how they could benefit or harm you. Quartz crystal, when programmed, has the ability to bring this information to your conscious awareness.

Crystals in Meditation

Quartz crystals are a wonderful help in meditation, as they have the ability to cut through confusion and help you tune into your Higher Self.

Take a cleansed crystal and "charge" it by centering yourself; then hold the crystal to your Third Eye. Program it with the purpose you wish to achieve. You can charge your crystal by focusing your desires into the luminous perfection of the crystal. When energy passes through a quartz crystal it becomes harmonized, and the natural balance is preserved. If the energies of the crystal user are not in balance, there will be a natural tendency of the crystal to correct and rebalance any energies that a person transmits through the crystal. This harmonizing ability will be enhanced if you speak to your crystal, telling it to balance your energies.

The natural tendency of the mineral kingdom is to create perfect balance and harmony. This elemental energy is largely directed by our Higher Will. We need only to tune into that Higher Self to live in harmony with ourselves, others, and the world.

Use your crystal to help you achieve an altered state of mind to access information which you otherwise wouldn't know. Use the crystal as a focusing device to reach a quiet meditative state. This altered state of awareness (sometimes called a trance state) allows you to delve deeply into the stored material of your subconscious to answer questions or to gain information. This trance state can also sensitize you to certain etheric vibrations so you can "see" the future or past. It is believed that sincere meditation with a crystal with positive, focused thoughts and firm mind control can solve all problems.

There is a technique in meditation which can be very helpful in benefiting from a crystal's energies. This is called closing the loop.

Seat yourself in a comfortable position while holding your crystal. To close the loop in meditation, hold the crystal between the right and left hand with the fingers of both hands touching the crystal for good electrical contact. The energy flows out from the right hand into the crystal which amplifies it and sends it into the left hand, onward up through the arms. Next, this energy goes into the hypothalamus and through the switchboard network, into the nervous system and throughout the whole body, and then down the right arm again and into the crystal for re-amplification and repeating. This is called closing the loop and gives you a strong energy flow in meditation.

After having a crystal for a period of time, many people find they can best meditate only while holding or wearing their crystal around the neck. If your crystal is suspended on a chain, ideally the best place to wear the crystal is low enough so that it hangs over the solar plexus or over the heart area. The solar plexus/heart chakra area is the most sensitive place for reception that we have at this state of our development in the human race.

An example of the great sensitivity of this area can be demonstrated by the feeling that when things go wrong, there is often a sinking feeling in the pit of your stomach. On the more pleasant side, when you have an overjoyed feeling, there is a great sense of exuberance. Often times people will make a comment like, "I feel it in my stomach." Or, "I have a good gut feeling about it." Or, "I have butterflies in my stomach." This is an open, intuitive part of the body, and an important acupressure area because it is so sensitive.

A valuable attitude to maintain while meditating is that it's important to overcome the lower ego self and realize why we're all here on Earth—to do the best we can to improve ourselves and to help each other.

Others Touching Your Crystal

Many people are concerned that someone who touches their crystal will "contaminate" it or instill it with negative vibrations.

As a general rule, with the average person there usually is not a problem. However, it is possible for someone who is very adept at programming and using crystal to put a message into a crystal that another person owned. It is possible that just by holding it in their hand for a few moments they could program their thoughts into the crystal, and those thoughts in turn would be rebroadcast to the inner body, inner consciousness, subconscious memory bank and all of the body cells of the person who owns the crystal. So, if someone was antagonistic to you and wanted to touch your crystal, you might want to cleanse it later and clear it because you might not trust their intentions.

For the average person who comes along, however, it doesn't matter in the slightest because it is highly unlikely that they would project anything into the crystal. In the first place, they probably wouldn't know how, nor would they have the concentration or the will power or the ability to accomplish any programming. All that would be likely to happen is that they would leave some fingerprints on the crystal which can be easily wiped off.

If there are any doubts, it is a simple matter to wipe the crystal clean. Just hold it in the sunlight and order it cleared of all memories, much like erasing a cassette tape. The crystal is then reprogrammed by talking to it while it is held in the hands. The programming enters the crystal's memory, both through verbalization and through body-cell contact with the hands.

Left Handedness and Crystal Working

There is an interesting phenomenon regarding crystal working and use of the left hand which is inherent in everyone's collective unconscious. This is very helpful to know when working with crystals.

Your left hand is always your receiving hand. Your right hand is always the powerful sending and giving hand—until you reach the stage in your self-development where you are able to balance yourself by explaining to your subconscious that you're a two-handed person with both a right and left hand, and they both will work equally as well. It does not matter if you are left handed or right handed. This concept holds true for everyone.

The reason for this programming is that for thousands of years every priestess/priest of oracles, every soothsayer, witch doctor, shaman, etc., supposedly was left handed. (This wasn't actually true, but they were supposed to be! If they were born naturally left handed, so much the better, and it did enhance their chances to become an honored person of divination.)

Their belief was that if one was born left handed, that meant the left side of them was open to receiving communications from the East. Everyone who went into a trance or sought to divine information was supposed to face south before going into a trance because this put the right hand, the strong hand, to the West where all Evil came from, and put the left hand to the east where all wisdom, knowledge, and love came from. God, the Sun and Mecca, etc. were all in the East.

Of course, early people didn't realize that the world was round and that everything came from the same place, but that was their reasoning for the importance of the left hand.

This belief was programmed into the subconscious (or collective unconscious) of everyone that the left side was different from the right side, and that the left side was the receiving side and the right side was the sending or giving side. Medical doctors will tell you that the nerves from the left hand go to a different place in the body than do the nerves from the right hand, so there is a physical difference as well as a metaphysical and subconscious difference.

For thousands of years this has been taught, and so it is in the subconscious race memory bank which automatically pushes little buttons in your nervous system saying that the left hand is what you're supposed to be using, instead of the right—in this case to feel the crystal's energies.

Summary of Programming Crystals

To realize the good in real quartz crystal psychic tools, it is necessary to program them. This is best achieved by holding the crystal in your right hand (the sending hand) and resting the right hand in the left hand (the receiving hand) to make a good electrical contact between them. This usage has nothing to do with being right handed or left handed, but is the way the sensory network operates. It is both traditional and factual. As the brain is divided into left brain and right brain depending on whether logic or intuition is used (which in turn is thought to be dependent on how the nerve network operates), so is the rest of the body also divided into left and right nerve systems.

The instructions are given to the crystal by talking to it. Consider the crystal to be a microphone which reacts to sound. The orders are given orally because a larger portion of the brain is put into action when thoughts are composed into auditory speech. Also, the crystal reacts to air vibrations as well as cell vibrations in the hand.

Another argument for using audible instructions instead of a silent message: Imagine you are in a room full of people and you are very thirsty. If you telepathically project that you want a glass of water, chances are slim that someone will pick up that request and actually go and fetch you a glass of water. However, if you make the request out loud, "Will someone please bring me a glass of water?" you will likely receive your water very rapidly. So it is with programming your crystal. Make your instructions or requests to it audibly. You will have much better success in obtaining what you ask for.

When held in contact with the hand, the crystal starts functioning on energies received from the physical body of the programmer. The crystal receives the vibrations, amplifies them and then broadcasts them outward in spiral waves. These amplifications are received by the sensory receptors in the left hand and relayed via the nerve system up the arm and into the brain to the hypothalamus gland.

The hypothalamus is a nexus of fibers lining the emotional center, the pituitary gland, the pleasure center, and the autonomic nervous system. It is the single most important part of the brain for maintaining homeostasis, and is a crucial link in any stress response.

This vital spot is often called the switchboard network of the body because all sympathetic and parasympathetic messages are routed through it. After receiving and decoding the meanings of these vibrations, the hypothalamus sends a record of them to the subconscious memory bank and then relays the vibrations through the body cells, and down to the right hand which is holding the crystal. The crystal receives the same message it has broadcasted, amplifies and rebroadcasts it again to start another circuit route for additional reinforcing.

On completion of the programming, it is best to keep the crystal close to your body by wearing it; or you could put it in your pocket. Somewhat like a tape recorder, the crystal keeps repeating its impressed vibrations indefinitely. This means that while you are busy at work, driving a car, dictating, using the telephone or doing other daily tasks, the crystal keeps reminding the body and its components of the original intent. This keeps the goal constantly active while your five senses are occupied with routine tasks. New programs can be added to the old as long as they are compatible.

To erase and cleanse a crystal, hold the crystal in the right hand as explained above and orally order the crystal to cleanse itself and erase all memories. This action is helped if the crystal is held in direct sunlight while talking to it. As previously stated, ultraviolet light passing directly through the crystal aids in a flushing action. After your crystal is erased, it is ready for a fresh, new program.

Note: These instructions are a modern revision of essential techniques handed down from classic, secret crystal ceremonies.

Photo by Frank Dorland

Examples of Carved, Polished Quartz

Crystal Handworking Tools

Photo by Frank Dorland

Chapter Six

HEALING AND CRYSTALS

Holistic healers and crystal users have demonstrated that crystals can be used to accelerate bone and wound healing, relieve pain, and help bring catastrophic illness into remission. Although it is said by some that there are no powers whatsoever in the crystal because it is a neutral object, its inner structure exhibits a state of perfection and balance. When a crystal is rounded, smoothed and polished to a form which augments its healing energies, and the human mind enters into a relationship with its structural perfection, the crystal emits a vibration which extends and amplifies the powers of the healer's mind. Like a laser, it radiates energy in a coherent, highly concentrated form, and this energy may be transmitted into objects or people at will.

When a person becomes emotionally distressed, a weakness forms in his subtle energy body, and disease may soon follow. (The subtle body is the invisible, ethereal substance which extends outward about five to eight inches from the physical body. It is invisible because it vibrates faster than the five physical senses can perceive. The subtle body is also called the astral body.) With a properly attuned crystal, a healer can help release negative patterns in this energy body, allowing the physical body to return to a state of wholeness. If a healer can attune his/her mind lovingly with a crystal, they become one with the Divine Mind which has imprinted its consciousness in the precise, geometrical form of that structure.

Healers use crystals because the crystal is the most perfectly organized state of matter existing in nature. It is precise, regular, and free from imperfections and impurities. Each person has the ability and hoped-for divinity to become a perfect being. Tuning into the perfection of the crystal can help us better comprehend this ultimate perfection and perhaps achieve it sooner and more easily.

One method which some healers recommend to use with a crystal in healing is to tune into the energy field of the crystal lovingly, projecting energy into the crystal, and resonating in harmony with the crystal. When you feel the crystal is charged, scan the subtle body or aura of the person, being sensitive to areas of obvious imbalance where healing energies should be directed. A good focal point is the heart chakra (see section on chakras for more information), where through visualization the problems felt intuitively can be brought to awareness. Then snap the crystal, like cracking a whip (by flicking the wrist and hand holding the crystal), releasing the negativity that was held in the subtle body.

The value of the crystal in healing is that it can take the feeling of love in the healer's heart and amplify it so that a concentrated stream of energy is emitted from the crystal for use in the healing process. This amplified energy field can help a person dislodge inhibitions which block the flow of their higher life energies.

The crystal works in much the same way that a laser does: it takes scattered rays of energy and focuses them. It makes the energy field coherent and unidirectional so that a tremendous force is generated. When used with love, the crystal unites the energies of the mind, and brings these energies into a pattern exactly fitting the life energies of the person seeking to be healed, then amplifies them for healing.

To use the crystal effectively, it is necessary to turn off the rational mind so that you may enter a right-brain meditative state and intuitively tune into the energies focused by the crystal transmitter.

There are many methods of crystal healing. Perhaps the best advice to follow is to listen to your own intuition. Use what works best for you. Practice on yourself and people close to you. Meditate and listen to what the voice within gently suggests to you.

Besides placing a crystal on the body where there's a pain or discomfort, a crystal can be held near the source of pain. In a clockwise motion, rotate the crystal and draw out the problem, shaking the crystal to get rid of the unwelcome vibrations. Single crystals can be placed near the soles of the feet and the palms of the hands to draw out blocks and balance energies.

Crystals are excellent biofeedback tools, and work well with creative visualization for mood and body changes. When feeling cold or chilled, hold a crystal in the left hand and draw in warmth. Do this in a color meditation with deep breathing, drawing in the energy, visualized as the color red through the left side of the body, in a circuit coursing through the body that releases it from the right side to the Earth or sky. Continue doing this for several minutes until you feel warm.

This works for cooling with blue as well. For cheering yourself up, draw yellow in this way; for calming, draw in violet or indigo. Body metabolism and heart rate can be lowered by this, an effect already familiar to those who have used deep breathing. Use crystals in visualizations, affirmations, and rituals to amplify and intensify what is being done.

Crystals can relieve pain almost magically. In another self-healing exercise, hold a crystal in the left hand, and feel the energy polarity build from it. (Polarity is a balance of positive and negative electrical energy. Because quartz crystal has a positive and negative charge on each of its faces, it interacts or sends a charge to the electrical field of the patient's body). Place the right hand gently on the pain area and hold it there. The pain is usually gone within half an hour.

A different method recommends that you set the crystal directly on the pain area. Hold it flat in the palm of your hand with the thumb, or grip it between the fingers, point downward. When you remove the crystal, the pain goes away from the body as well.

Size of Healing Crystal

The general size of crystal to use for healing is whatever fits comfortably in the palm of your hand, or, what you are attracted to use. The size of the crystal is not so important as the feel or essence of it. It isn't necessary to have a large piece of quartz; a small piece will suffice in most cases, especially if it is clear, electronic quartz.

The prime importance is the proper application of the human mind. The crystal (small or large) is a valuable tool to be used in harmony with the mind because the crystal is a *reflecting amplifier* to communicate healing messages to the body cells, the immune system, and the subconscious "commander" residing in the patient.

Crystal Healing Techniques

There are many ways to use quartz crystals in healing. Each healer has their own favorite method which works best for them. Two common techniques which have been shown to be useful are using the crystal for gentle *acupres-*

sure work with the more pointed pieces, and for *massage* with the more rounded parts.

One method which has proven to be successful is this: Hold your healing crystal in your left hand. Make a pass with the crystal held one to two inches above the body, slightly above the physical body but not touching it, until you locate the problem spot. The crystal will magnify any changes in the magnetic flux field (the aura or subtle body) so that you can easily be aware of them. You will notice a different feeling in that location.

Often, where the patient's pain seems to be is not where the root of the problem is. For example, if the patient says, "My neck hurts," you may find that the crystal "feels" different on the lower back or somewhere else on the body other than the sore neck. Where the energy change is felt from the crystal is really where the seat of the problem lies. The crystal should then be applied to the physical body.

Next, change the crystal to your right hand and gently vibrate it, touching the body in the spot where the problem exists. Do this gently otherwise you could bruise the skin or tissue. There should not be any pain involved from the pressure of the crystal on the person receiving the healing. An actual electrical charge occurs with crystal healing. A good healer is automatically aware of this charge and what is occurring in the patient's body, and he/she will receive biofeedback from their body and will administer healing as long as it feels right. However, to

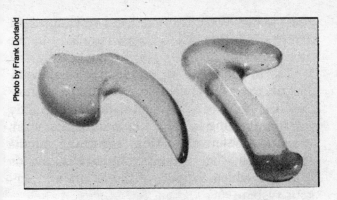

Carved Quartz Crystal for Healing

gently pointed end for acupressure   rounded end for massage

put a jolt of electricity even as delicate as that of quartz crystal's energy into the physical body results in a definite reaction on that body, so no more than four or five minutes over the whole body is required for this portion of the treatment. Just spend a few seconds on each sore spot, and then gently vibrate the crystal on it.

The reason that you vibrate the crystal is to gain the attention of the body cells. If you merely place the point there and hold it motionless, the body cells record it as just a pressure and don't pay much attention to it—they are used to pressure on the body all day long. However, when you vibrate the crystal and focus your own healing energy through the crystal, you send the message to the patient's body cells that "Here I come. I am a powerful healer and I'm going to heal you." The body cells will pay attention to this message and they will relay it to the subconscious.

The important thing is to get the message to the subconscious and to the body cells, because they are the ones who control the healing. The subconscious memory bank controls your immune system, digestion, breathing, blood pressure, etc.—all of the involuntary actions of the body which are part of the autonomic nervous system.

When enough time has elapsed so that the problem area has been energized, go on to healing other parts of the body which have indicated resistance, heat, or coolness. For a headache, rub the flat portion of the crystal on the forehead.

Lastly, smooth out the aura by using brushing hand movements as if you were brushing off invisible dust and debris, about three to five inches above the body.

Remember, this acupressure part of the treatment needn't take more than four or five minutes, because the person needing the healing must be kept comfortable at all times. To finish up the treatment with massage and laying on of hands (see section on laying on of hands for more information) is also very healing and soothing, and may take as long as the healer wishes.

One more important aspect of healing is trust and faith. Healing occurs when there is trust—trust of the patient's body cells that someone is helping them.

If a patient goes to see and doctor and they do not have a good rapport, or if the patient doesn't like or trust the doctor, the MD will send the patient to another specialist. Unless the patient trusts and likes the doctor, there will not be much of a healing effect.

If a doctor or healer can convince the patient that they have special abilities to heal, immediately the immune system is calmed. The message goes to the body that "I can relax. S/he is going to cure me." The charming countenance and good intentions of the physician often are as important as the medications prescribed. The placebo effect is extremely important.

How and Why Quartz Crystal Helps the Body

As previously stated, the body has electrical currents running through it. This is one of the main reasons why quartz crystal is such an effective healer—it generates a flow of electricity from the hand and body of the healer, amplifies it, and steps up this flow into the body of the patient. Most diseases and problems with the physical body are the results of blocked energy.

Quartz crystal actually helps in healing in two ways. One is by transferring this gentle electrical energy into the body where it is needed. The other way is by amplifying positive programming to get well and stay well. Using a crystal in this manner inspires the body to boost the immune system to achieve a better state of health.

Recent research has shown that the body's immune system can be turned on and off by external influences—mind over matter. One's mental attitude is supremely important in healing. Metaphysical healers have said this for years.

At the core of the immune system is interferon. The body produces interferon to fight disease and infection. Interferon is a cellular protein which is produced in response to, and acts to prevent replication of, an infectious viral form within an infected cell. Doctors who experimented on patients have found that interferon increases under certain stimuli. The level of interferon in the body was determined by blood tests.

Clinical studies have shown that the body can produce interferon, the body's natural immunizing agent, by using mind control and will power. The use of quartz crystal in healing and self-healing greatly amplifies the brain's ability to influence mind over matter.

Brain Wave Levels and Healing

For effective healing of yourself and others, it is necessary to understand the different levels upon which the brain functions, and how to enter the state which is most conducive to healing.

When a person entertains negative thoughts, and then as a consequence has a negative emotional and physical response, a change occurs in the brain wave frequency. By understanding the brain and its electrical nature, you will have a better idea of how to consciously change your life and enhance healing in your life and in the lives of others. When working with quartz crystal, remember that the crystal will amplify all thoughts you have, so it's important to eliminate negative thought patterns as much as possible.

The human brain generates electricity. A person has approximately twenty million brain cells, all of which are capable of carrying an electrical charge. Each of these brain cells has an axon which functions as an electrical charge, an axon which functions as a neuron which carries the electronic impulses *outward* from the cell, and between one and twelve transmitters called dendrites. This configuration allows for trillions of interconnections between brain cells. (A dendrite is a protoplasmic extension of a neuron—usually several to each neuron. Their function is to establish the chemical and electrical communication systems to other nerve cells.

When the brain cells are at rest, each cell has a potential electrical force called voltage. When these same cells become active, they release the potential energy and an electrical current is generated which carries the message to other nerve cells.

The electrical impulses generated by the brain and the patterns they create have been studied and analyzed by an instrument called an electroencephalograph (EEG). With the EEG, the general patterns of brain wave rhythms common to all people have been identified, along with their relationship to different states of consciousness.

Brain wave rhythms have been grouped into four major categories: alpha, beta, theta and delta. The patterns or rhythms caused by the brain's electrical activity are measured in cycles per second (cps). It is generally agreed that waves of about 14 cps and higher are beta waves; those of about 7 to 14 cps are alpha, those of 4 to 7 cps are theta, and those of 4 cps and less are delta waves.

Brainwave patterns vary within each category, and this is why they cannot be characterized as one specific frequency and must be defined in ranges of frequency. Also, different factors can alter the readings of the EEG. For instance, the placement of the recording electrodes can give different records of brainwave activity depending on where they are placed—a different reading will result from their being placed on the frontal or pre-central areas of the skull.

An important discovery has come out of biofeedback research. Intense concentrated attention can be present in an individual even when his EEG pattern shows him to be in a state identified as relaxed and inattentive. The individual may exert a great deal of energy in an effort to control his alpha or theta activity, yet his EEG recording will show no sign of unusual effort, no indication that he/she is alert and concentrating on events in the external environment. Perhaps this is the case of healers who are relaxed, yet concentrating on sending energy to their patients.

DELTA: The slowest brain wave frequency is called delta. It operates between 0 and 4 cycles per second. Delta appears only during the deepest levels of sleep, in a coma, or while a person is under anesthesia.

THETA: Theta waves are the next frequency, operating between 4 and 7 cycles per second. They are rarely found in a normal, waking individual. Theta waves are associated with drowsiness, the assimilation of new information and creativity. They are present most frequently while a person is deeply relaxed or daydreaming. Although theta is normally associated with sleeping, it will suddenly appear during periods of insight or inspiration, and during deep, healing meditation. The best time to program crystals to effect change and healing is during theta brain wave activity. Some people feel that the theta level is also the most creative state.

ALPHA: The alpha level has a frequency of between 7 and 14 cycles per second. This is a desirable level to maintain, as it is very beneficial to the body and mind. Renewing and self-healing is accelerated, and stress is minimized. When an individual is in a meditative state, the alpha level is the most common state of brain wave activity.

BETA: All brain wave frequencies above 14 cps are under the category of beta. This level of activity is associated with alert, rational, analytical behavior. The beta level is regarded as a stress state. It has been discovered that negative feelings such as guilt, anger, jealousy, etc. do not occur at lower brain wave frequencies. Unfortunately, most adults spend about 80 per cent of their waking time in the beta level, while prepuberty children spend 80 per cent of their waking time in the alpha state.

It is now common knowledge that *attitude* is very important for the healing process. It can greatly accelerate healing, which is enhanced at the alpha level. It is generally accepted that people who function at the alpha level have stronger immune systems and heal more quickly. They have more control over pain and can more easily lessen its deleterious effects than those who function primarily at the beta level. Those who can achieve the alpha level easily are able to control involuntary physical responses such as blood pressure, heart rate, body temperature and even bleeding. Being in the alpha state and projecting healing or energy with a crystal go hand in hand.

Placebo Effect in Healing

The American Heritage Dictionary defines placebo as "a substance containing no medication and given merely to humor a patient." In unorthodox healing, many methods of healing and types of cures were attributed to the placebo effect of mind over matter. Now research has shown that the placebo effect and the power of the mind are essential factors in physical healing regardless of the healing procedures used.

Throughout history, people have been cured by many unusual drugs and remedies, from swallowing snake oil or tapeworms to the application of leeches to ingesting rhinoceros horn or crocodile dung. Most of these remedies were inherently worthless except that the patient believed these things would make him/her well, and so they did. The placebo effect phenomenon even extends to surgery. An experiment was conducted in the late 50's and 60's with chest surgery for angina pectoris (oppressive chest pain). Half of a group of patients had actual surgery whereby the mammary arteries were tied off, a procedure which was thought to reduce pain. Half the patients were anesthetized and had a small incision made, then sutured in their chests. When they were awakened they were told the operation had been successful. Later, when all the results from the study were taken, physicians found that the patients who were given the simulated operation fared better than those who actually underwent surgery.

All of this goes to show the *importance of the mind* in healing. It has emerged as the principal factor in healing and in maintaining good health.

Chakras

Understanding of the chakras is of primary importance to being a good healer, and to effectively work with your own energies. The word chakra means wheel in Sanskrit. It refers to one of the seven major centers of spiritual energy in the human body, according to yoga philosophy. Actually, there are hundreds of minor chakras or energy points in the body. They occur wherever there are nerve endings coming together, and are the pressure points in the body. The explanation here will concentrate on the seven major centers that most people are familiar with.

Chakras are psychic centers located within the cerebrospinal system. By understanding and utilizing the energies of the chakras, we will ultimately reach an enlightened state of being and become all that we are capable of being. The chakras are centers of activity of a subtle, vital force called *prana*. They are interrelated with the parasympathetic, sympathetic and autonomous nervous systems, and through these systems, the physical body is joined to the chakras.

Clairvoyants perceive chakras as colorful wheels or flowers with a hub in the center. Each is attuned to the spine by an invisible tendril, much like the stalk of a flower (see illustration). The chakra system runs up and down the length of the spine, from the bottom or root chakra to the top of the head or crown chakra.

Each chakra has a counterpart in the astral body, but draws energy in from the etheric body. They appear as a depression in the etheric body, and through them the energy flows in. This ener-

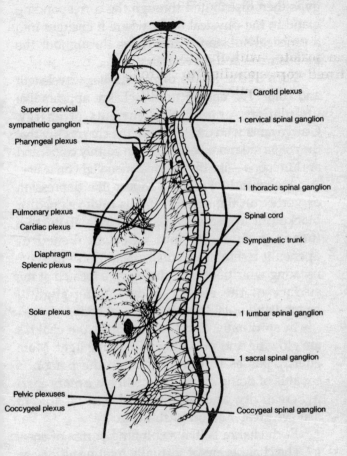

Carotid plexus

Superior cervical
sympathetic ganglion

1 cervical spinal ganglion

Pharyngeal plexus

1 thoracic spinal ganglion

Pulmonary plexus

Spinal cord

Cardiac plexus

Sympathetic trunk

Diaphragm

Splenic plexus

Solar plexus

1 lumbar spinal ganglion

1 sacral spinal ganglion

Pelvic plexuses

Coccygeal plexus

Coccygeal spinal ganglion

THE CHAKRAS AND THE NERVOUS SYSTEM

from *The Chakras* by C.W. Leadbeater,
Theosophical Publishing House,
Wheaton, IL. 1980 edition

gy is then distributed through the corresponding gland in the physical body, where it changes into a physical substance and flows throughout the bloodstream and nervous system.

Each chakra has its own particular wavelength and color. The energy from a chakra appears like undulations or radiating currents of energy. Clairvoyants who can really see the chakras say that the colors are more beautiful than earthly colors and are luminous—like the gleam of moonlight on water.

A chakra looks like a saucer-like depression or vortex on the surface. If it is undeveloped or blocked, it appears as a small circle about two inches in diameter, but when awakened or opened it resembles a small, blazing sun-like coruscating whirlpool. The chakras are located at the surface of the etheric body—going slightly beyond and outside of the physical body.

In spiritually developed people, the chakras are glowing and pulsating with living light. More energy passes through them, so the person is capable of doing more. They will have more energy, creativity, and vitality, and will be able to achieve greater accomplishments.

Each chakra is like a spinning vortex of energy. The chakras are perpetually rotating as long as life exists. As the chakras spin and release energy into the body, these energies saturate the body cells with life and vitality. This energy or vitality then radiates out through the force field (the aura) that surrounds the body. The inrush of ener-

gy from the chakras is necessary for the existence
of the physical body.

If the chakras are not balanced, or if the energies are blocked, the basic life force will be
slowed down. The individual may feel listless,
tired, out of sorts with the world, and depressed.
Not only will physical bodily functions be affected so diseases may manifest, but the thought
processes and the mind may also be affected. A
negative attitude, fear, doubt, etc., may preoccu-
py the individual.

If the chakras are opened too much, a person
could literally short circuit themselves with too
much energy going through the body. You have
probably seen examples of someone like this
who burns the candle at both ends and seems to
be constantly hyper and on the go. Sooner or
later, the person will burn out and will not know
how to maintain such a high-energy state.

The reason why quartz crystal is so effective
in balancing and energizing chakras is because
of electricity. Electronic quartz responds to the
electricity that is coursing through the body, and
if this energy is sluggish, the undeviating electrical vibrations of the quartz will help to harmonize, balance and stimulate these energies. Some
healers like to place a piece of quartz directly on
the chakra, while others prefer twirling or rotating the crystal a couple of inches above the
chakra, not touching the physical body (the
direction in which you rotate the crystal is less

important than the intent. Go with what intuitively feels right to you). You can also place a crystal anywhere on the body where there's pain or discomfort.

THE FIRST CHAKRA is known as the root chakra. The Sanskrit name is *Muladhara,* meaning "root/base." Its location is at the base of the spine along the first three vertebrae. Its function is for survival and grounding. If this chakra is blocked, an individual can feel fearful, anxious, angry and frustrated. Clues to this will be tightened jaws and fists, and sometimes violent behavior based on insecurity. Obvious physical malfunctions may manifest in problems such as obesity, hemorrhoids, constipation, sciatica, degenerative arthritis, anorexia nervosa, or knee troubles.

When the chakra is open and balanced, the individual will be in balance and will act wisely and with moderation. When working with this chakra, it is helpful to visualize the color red or black surrounding the area.

THE SECOND CHAKRA is known as the sex chakra, or in some philosophies the spleen chakra. The Sanskrit name is *Svadhisthana,* meaning "one's own place." Its location is on the genitals along the first lumbar vertebra. The function for this chakra is desire, pleasure, sexuality, and procreation. If this chakra is blocked, a person may be restless and confused. Physical malfunctions may result in problems such as impotence, frigidity, uterine, bladder or kidney trouble, or a stiff lower back.

When the chakra is functioning properly, the individual will feel emotional gratification, courage, and attraction to the opposite sex. The color to visualize for strengthening or balancing this chakra is orange.

THE THIRD CHAKRA is known as the solar plexus chakra. The Sanskrit name is *Manipura*, meaning "jewel city." Its location is in the region of the navel along the 8th thoracic vertebra. The function for this chakra deals with developing the ego. The solar plexus is the seat of feelings and emotions. It deals with will and power. When this chakra is balanced, the need to feel important and achieve outside material identity in the world is transformed into true contentment and faith that what is occurring now is what is really needed at the moment for complete spiritual growth. An open solar plexus chakra is shown by acts of charity and selfless service. The color to visualize while healing this chakra is golden yellow.

THE FOURTH CHAKRA is known as the heart chakra. The Sanskrit name is *Anahata*, meaning "unstruck sound." It is located in the center of the chest in the heart or cardiac area, along the 8th cervical vertebra. The function of the heart chakra deals with love. It is one of the most important of the body centers. It is said that if this chakra is open, all the other chakras will come into alignment with it. The heart extends its circulation to the entire body, and all systems and tissues are nurtured with vital force. If the heart chakra energies are blocked, physical symptoms such as asthma, high blood pressure, heart disease, or lung disease may result. An open and balanced heart chakra demonstrates compassion, and the ability to see and feel through the eyes and heart of the other person without losing or betraying oneself. The color to visualize in balancing this chakra is green.

THE FIFTH CHAKRA is known as the throat chakra. The Sanskrit name is *Vishudda*, meaning "with purity." It is located in the laryngeal area, at the base of the throat along the 3rd cervical vertebra. The function for this chakra deals with communication, creativity and speaking higher truths. When this chakra is balanced, the cosmic laws of God and life are understood and lived. One will have calmness, serenity, purity, a melodious voice and a good command of speech. If it is blocked, physical symptoms may manifest such as a sore or stiff neck, colds, thyroid or hearing problems. The color to visualize for healing problems with this chakra is blue.

THE SIXTH CHAKRA is known as the brow or third eye chakra. The Sanskrit name is *Ajna*, meaning "command center." It is located in the middle of the forehead in the area between the eyebrows, along the 1st cervical vertebra.

The function of this chakra is clear seeing and intuitive sight. When this chakra is balanced and opened, the third eye is able to see clairvoyantly. Intuitive knowledge is available for help in making decisions. If the brow chakra is blocked, malfunctions such as blindness, headaches, nightmares, eyestrain or blurred visions may result. The color to visualize for balancing this chakra is indigo.

THE SEVENTH CHAKRA is known as the crown chakra. The Sanskrit name is *Sahasrara* and means thousand-petaled lotus. It is located at the top of the head. The function of this chakra deals with perfect understanding and having a higher state of consciousness. Opening this chakra results in bringing through and manifesting higher wisdom and Divine Light and love. If this chakra's energies are blocked, physical symptoms such as depression, alienation, confusion, boredom, apathy or an inability to learn or comprehend will sometimes be apparent. This chakra represents the highest level of attainment for understanding cosmic bliss while in the physical body, and is the last to be developed in an individual. The color to visualize while balancing this chakra is white or clear, representing all colors. Sometimes violet is used.

Some healers like to begin a healing session by checking out or scanning each chakra with a crystal held in their left hand. This increases their sensitivity to feeling blocks or cold spots on these energy points. Stroke the area an inch or two above the body from the top of the head down to the toes. Feel for heat, tingling, cold, or resistance. Either gently massaging the crystal on the chakra or rotating it gently above the area can be effective in opening blockages. Sometimes it is helpful to place a crystal or gemstone on each of the seven main chakra points. Or, place a crystal anywhere on the body where there is pain or discomfort.

When you are finished with the healing session, again lightly brush the body down the front and the back to cleanse and seal the chakras from any undesirable energy which may interfere with your healing. A healing like this may be likened to psychic surgery, and the energy centers may be sensitive to harsh vibrations for a short while and so should be protected from unwanted influences.

The Aura and Healing

Every living thing has an emanation of light surrounding it. Plants, animals, humans—all life gives off a radiation of energy which is called the aura. This aura is divided into three main layers: the first or permanent layer, the second or emotional layer, and the third or spiritual layer. It looks like an egg-shaped bubble of colored light emanating from the physical body.

It has long been accepted by science and metaphysics that a slow, direct current of electrical energy flows through the nervous system of the body. This flow is affected or blocked by certain conditions such as illness, injury and even anxiety. This pattern can even be affected by external electromagnetic force fields (which is why it is a wise idea to wear a programmed crystal to repel negative effects of these force fields).

Many things are believed to affect the color of the aura such as the state of health, emotional condition, age, thought patterns, etc. We are always aware of the auras of other people around us, mainly through our feelings. A mother is quick to sense if a child is ill, even if there are no apparent manifestations of disease. She is able to do this by sensing the auric field.

It is very important for healers to be familiar with the aura and the problems with it which are readily apparent when good health is lacking. By knowing how to intuit or "see" the aura, it is easy to tune into where problems will later manifest in the physical body, because disease first appears in the outer layers of the aura and only manifests in the physical body as a last resort to warn the body that all is not as it should be, that one's thought patterns have given way to disease (dis-ease).

An average aura usually extends out from the body about three feet. In more evolved or developed souls, it may extend out several feet. It is said that the aura of Buddha could be felt for over a mile.

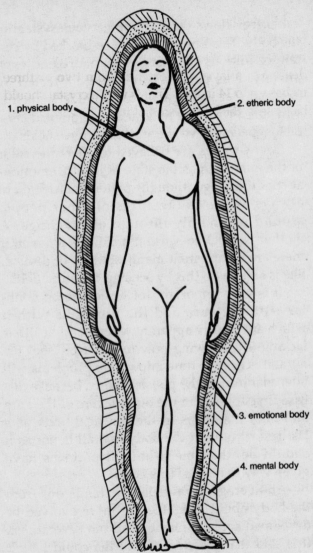

1. physical body
2. etheric body
3. emotional body
4. mental body

Outline of Body Showing Different Layers

While the aura may radiate outward to distances of even 10 or 12 feet, the major active portion (at least 90 per cent) of the energy is condensed in a layer that extends from two or three inches up to 14 inches. That is why a crystal should be close to the body—so that it can be electronically activated by the auric flow of energy.

The color of a person's aura may change with the thought patterns he is entertaining. Following are some descriptions of what an aura may look like depending on the temperament: If someone was depressed, his aura might look gray, dim and dismal. If the same individual picked himself up and thought about his love for humanity, his aura would change to a deep red. Then if remorse should be felt, the aura would become a dark blue. Feelings of jealousy would show dark green.

A person who is good at meditating may in the course of half an hour change the colors of his aura three or four or five times as he moves his awareness from the deeper tones of intellectual reasoning to the more brilliant hues of spiritual realms. The aura may then take on shades of light blue and light yellow interlaced with white.

We all sense others' auras every day. Most of the time we tune this information out, but it is there for our use. It is easy to sense that someone is not feeling well, or that they are disturbed. We know this because we are sensing or seeing with our third eye. A quartz crystal can help you to see auras more clearly by boosting your intuitive feelings.

The human aura was made visible to the scientific world when Dr. Walter Kilner in the 1870's developed a special coal tar dye made of dycyanin which enabled anyone to see an aura. By painting this dye on a lens, one could see the ultraviolet spectrum. With his "Kilner screen," he could observe the aura or energy field around a body. Although he published his findings in 1912, the medical and scientific establishment was not yet ready for his discovery.

In 1939 the Soviet engineer Semyon Kirlian discovered through experimentation how to produce a photograph that showed a luminescent form around the fingers of his hand. This was the living energy radiating from his hand. Kirlian photography was thus born. Kirlian, together with his wife Valentina, discovered this useful method of photography which became a useful tool in the hands of psychic researchers, and has since been very helpful for scientific research as well.

Use of Kirlian photography has been used to heal certain ailments simply by "radiating" the body for brief periods of time. It has also been used to scan the body for potential problems. With modern color film and high resolution, even minor changes in the body become apparent to researchers, such as effects of music or certain stimuli on the body.

The aura has undergone different interpretations by psychics. Some call it a spirit form, which has been seen after the death of the physical body. This spirit form can often be seen around the area where it made its home when it was alive. Another interpretation is that the aura is the astral body which is capable of dislodging itself from the physical body and traveling to distant locales. Upon returning, if memory is intact, the person to whom the astral body belongs will have remembrances of what was seen and done.

The most common understanding is that the aura is an electromagnetic force field surrounding the physical body. Whatever the aura is, it presents a vividly colored book enabling one to read a person's past, present and future. It contains an eloquent language which is richly descriptive and, if understood, can be immeasurably helpful to the person in this life's journey.

A good healer will scan the aura to pinpoint where problem areas are. Again, as with the chakras, where there are different or blocked energy flows are places which need to be stimulated with energy from the crystal held in the hand of the healer.

Color Healing

Color plays a very important part in our lives. Our clothes, the color of our rooms, and our total environment all affect our disposition. Scientific studies have determined that students taking tests in rooms painted red exhibited more anxiety. When they took tests in rooms painted blue, the students reported feeling more depression. The color of our surroundings certainly can affect performance.

Previously, schools used to have drab walls—all were grey or beige. Now, especially in the preschools and elementary schools, the walls are painted in bright, cheery colors. Students are happier and they function better with these pleasing colors. Hospitals too have started painting their walls in more pleasant colors, and patients respond to the healing hues. Medical facilities seek to cheer those patients who are sluggish and low by the use of reds, pinks and orange (warm colors) in their environment; for those who are overexcited and hyper, blues and greens are used (cool colors).

Nowhere is color more evident than in Nature. The trees, flowers, sky, plants, birds, animals, etc. all offer their exciting array of color for our eyes to feast upon. It would be a drab world indeed if we did not have the pleasurable experience of seeing color. It is healing and tranquil just to be outside in the summer and to see the green grass and green leaves on the trees.

Healers have long known about the effect of color to initiate a change toward wellness in their patients, and have known how to use and visualize color in their healing techniques. Color is extremely important in healing. It is important to the emotional body.

We have been programmed to believe that a certain color will have a special effect on us. Different nationalities and cultures have different "programs" built into the subconscious of their people, so a given color will have a different effect on people from other cultural backgrounds.

A unique characteristic of color is its ability to create a response without our conscious awareness. You do not even need to know that it is healing or how it heals you; just by the very vibration of its wavelength, a given color benefits you.

Color healing works first on the emotional body. There is more to each of us than just the physical body, and more than is apparent to the five senses. On the outside of the physical body, extending for about two inches and interpenetrating the physical body, is the etheric body. This can sometimes be seen as a faint, grayish outline. On the outside of the etheric body is found the emotional body. As expected, this is where our feelings and emotions are stored. The outside casing of these areas is called the mental body.

Disease starts in the outer, invisible body. When an illness manifests, it has penetrated through all of the protective layers and finally reached the physical body.

The visible spectrum of light apparent to the human eye can be broken up into a prism of seven main colors. These are the same colors as are seen in the rainbow, going from the longest to the shortest wavelength: red, orange, yellow, green, blue, indigo and violet. These same colors are generally associated with the seven main chakras, with red corresponding to the root chakra and violet (or sometimes clear) corresponding to the crown chakra.

There are traditional associations regarding color and healing. Specific colors for certain kinds of ailments have been used throughout history. These techniques consisted of such things as drinking colored water, using colored cloths placed on the part of the body which needed healing, viewing colored objects and breathing in color. Holding or wearing a colored gemstone or crystal is a very important method of color healing still used today. Color meditations wherein an individual will visualize him/herself surrounded by the needed color have been effective too. More recently, healers have the patient sit or lie down in front of colored lights or colored lenses for a period of time.

Fortunately for us, on the Earth there are members of the mineral and plant kingdom which are beneficial to humankind. These helpful stones and plants affect us mainly through color. The interaction of gemstones' energy can sometimes be felt as heat or a similar type of energy to a sensitive and aware person. The most beneficial plants are those with fragrant blossoms and fruit, most of which have bright, beautiful colors that attract us to them in the first place.

Following is a list of some of the more common associations for using color in certain treatments, both physical and psychological:

RED: Physical ailments treated by using the color red include problems with the bloodstream such as anemia, low vitality, poor circulation. Red vitalizes the physical body and is a powerful stimulant. It should not be used for people who are nervous and irritable. Too much red results in feelings of anger, frustration, lust, violence, destruction, etc. The positive understanding of red results in strength, power, initiative and honor.

ORANGE: This color can be used to treat problems with the spleen, lungs and pancreas. It is helpful for muscle spasms and cramps. Orange has an enlivening effect and provides energy to deal with life's situations. This color has been used to treat asthma, bronchitis, gall stones and menstrual problems. Negative aspects of orange energy result in aggressiveness, uncooperative behavior, inferiority or superiority complexes and procrastination. Positive use of orange results in illumination, confidence, intellect, inventiveness and self-motivation.

YELLOW: The stomach area is successfully treated by yellow. Problems such as constipation, indigestion, bile flow, diabetes, flatulence, heartburn and skin troubles are helped with yellow. Yellow activates the motor nerves and generates energy in the muscles, plus it helps to eliminate depression. Yellow is helpful for the intellect. Negative use of yellow results in criticism, over-indulgence, stubbornness, cowardice and being judgemental. Positive use results in joy, mental discrimination, organization, attention to detail and discipline.

GREEN: This color is used to treat problems in the heart area. It has been used for blood pressure, ulcers, and headaches. Green is a very important color as it lies midway between the two ends of the color spectrum. Its function is one of balancing, with harmony, peace and serenity resulting from it. Green quiets and refreshens the mind and body. Qualities of green lend determination, efficiency and conscientious attention to detail. Misuse of green energy results in jealousy, envy, stinginess and greed. Wise use of this energy lends enthusiasm, hope, sharing, growth and expansion.

BLUE: The color blue is helpful in treating all throat ailments such as sore throat, hoarseness, goiter, fevers and laryngitis. Blue is very peaceful and relaxing and has the capability to draw one out of the physical world and up into the spiritual world. The negative qualities of blue are shown in depression, self-pity, fear, coldness, detachment and indifference, while the positive use of blue results in wisdom, gentleness, trust, understanding and forgiveness.

INDIGO: This deep blue/purple color is helpful for problems commonly associated with the head, such as eye, ear and nose troubles. It is also beneficial in the treatment of certain nervous and mental disorders. It works on the parathyroid glands, but depresses the thyroid. Indigo helps inflamed eyes and ears, and has been used to treat pneumonia. Misuse of the energy from indigo can result in pride, separateness, conceit, gossip, deceit and irritability. The positive aspect of indigo helps bring about qualities like unity, calm, balance, humanitarian and world service, synthesis and aspiration. Indigo also acts as a catalyst.

VIOLET: This color is best used for treatment of nervous and mental disorders. It helps purify the blood and stops the growth of tumors. It controls the pituitary gland. Insomnia is helped with violet, as are eye troubles. It has been used to treat sciatica, meningitis and epilepsy. Misuse of the violet color can result in obsession, martyrdom, injustice, intolerance and restriction. Wise use of the qualities of violet reveals mercy, devotion, loyalty, idealism, wisdom and grace. Violet represents the highest element in human nature.

Use of these colors can enhance all forms of healing. When possible, it is helpful to have several different colored quartz crystals, but it is almost as effective to use a clear quartz and visualize the color needed in treatment. The quartz crystal will amplify the thought forms and give more energy to the focus of attention.

It is believed that the main attribute of quartz crystal is its amazing ability to transfer electrical vibrations, and that color is a secondary attribute. However, many healers prefer using colored quartz in their work. Some have a healing tool in every color in which quartz is available.

Some professional healers have found that they obtain better results from crystals because of a patient's belief. For example, when using a green crystal they explain that the green is like springtime when all the grass begins to grow. The green is a healing, new, fresh and invigorating color, and that is what will be put into the patient's body cells. Other people who are hyped up or nervous find that a blue crystal brings them a tranquil effect, and represents serenity and peacefulness. If it works more successfully, then it is better to use a colored quartz crystal healing tool.

Preprogramming a Crystal for Healing

When using a crystal for healing, the question has come up whether or not it is necessary to program the crystal beforehand.

It is effective to just use the crystal and program it as you are using it from your body cells and the body cells of the patient. It is not necessary to have previously programmed it, but as you are using the crystal, think of the healing you are trying to accomplish. Have a focus of your good intentions.

When you preprogram a crystal, that would be for your own personal benefit and use. For example, you program a crystal, then you put it on a chain around your neck so that it is next to your skin.

At some later time, suppose you are driving down the street, or you're typing or talking on the phone. The crystal is close to your heart chakra and the solar plexus center. It is sending the vibrations that you have programmed into it to all your body cells, and this keeps you on an even keel, in the direction you want to go. It keeps reminding you what your prime purpose is so you don't deviate, you don't get lost in the confusion of everyday life. You stay on the right track. That is the idea behind programming a crystal. You tell it what you want yourself to be doing—what your fulfillment in life is going to be, and then it keeps you directed on that energy level.

Now if someone comes to you for healing, and you as a healer are doing laying on of hands or acupressure with a healing crystal, you don't want to put all these other messages meant for you into their body cells. All you want to place in their body cells are the orders for the body cells to balance the body chemistry; you also want to find the problems and send all the necessary white or red corpuscles, or elso do whatever needs to be done for healing.

There are some healers who use a technique of inviting the cell consciousness of the patient to come into the body of the healer by saying, "You don't know what you're doing, obviously, so come into my body and see how to operate properly. Look at how well my body is doing. Use me as an example. See that I am well balanced and well organized, and that you are confused. Come on in and look at what I'm doing."

You tell that to the other body cells silently through the crystal. It could be upsetting to the patient to hear them audibly, so it is best to send this message silently. Audibly you can say comforting words of healing.

Many times women make the best healers because of their unique ability to use both sides of the brain at the same time. With their left brain they can audibly comfort the patient with reassuring words of healing, while with the other half of their mind they can give directions through their body cells. They are giving very absolute, serious, definite and fundamental directions to the body cells of the patient, and telling him/her exactly what to do.

(Obviously, men who are in tune with the compassionate healing force will be successful too.)

It is effective to "threaten" the patient's body cells/lower self and scold the lower consciousness. It is also effective to begin by establishing your authority as a healer to the body cells of the patient (remember, all of this is done silently). Mentally say, "I am the most powerful healer in the entire universe and you are going to have to answer to me every day until you do what you are *supposed* to do and heal this body. You must do as I tell you, and do it right now. Now get to work and correct the condition of _____."

Granted, the energy for healing comes from the universe, the Divine Source, and the healer draws on this energy and channels it through his/her body into the body of the patient. But as a healer you must have perfect confidence in yourself. You must be in command and you must give the orders. The power does come from God/the Source, but we must remember that we are all gods in the making.

We are each an extension of God. We are bringing this Divine Light/energy down to the human level to use in order to become the perfected beings we were meant to be.

Jesus said, "All these things and more can you do." Each person has the capability to be a healer—to heal the self and others.

A good healer must take control. A healer has to be dominant and in charge and tell all the body cells what to do to overcome their diseased status. It is very important for the healer to believe in himself. If the healer is not serious, the body cells of the patient won't believe in him, and the healing will not be effective.

You must have the faith in yourself and the faith that this power is coming through God and through Universal Intelligence, and that you are using it properly and for the benefit of the person you are working with. You must be firm, and in direct command at all times. The patient's body has likely been programmed for failure and illness, and you must "shake it up" to get its attention.

Self-Healing with a Crystal

Quartz crystal can be an excellent help for self-healing and it can be used in several different ways to bring about healing.

First of all, it is a good idea to ask the crystal for information and advice on what is really wrong. You can do this through programming, by saying such things as, "Where am I really sick? I know where I hurt, and the doctor said I have this particular disease. Is this true?" Ask the quartz crystal for answers.

You could also use a crystal as a pendulum because the pendulum provides you with a cell-conscious answer which works through the subconscious memory bank. If you are unfamiliar with the use of a pendulum, ask it to explain to you what it thinks is wrong. Then if you have something drastically wrong you could treat it by using your own mental mind as a holistic type of healing. Do not hesitate to also have the best medical treatment at the same time, because the two methods, the medical healing and the mental healing, complement each other. One should not give up medical healing, because it has aspects that are very necessary and useful.

However, it is less effective to undergo medical healing without mental healing. It is difficult to get well in the hospital without the love of a caring nurse or family member or friend, unless one is a very strong-willed person and can say, "I don't give a darn what they say, I'm getting out of this place."

In self-healing, you can either wear a crystal that you have programmed for healing, or you can also use it to massage yourself on the pressure points. By this massage, you would amplify the orders you are sending to the body cells.

The programmed crystal responds to your own mental powers. The only thing that will activate the crystal is your programming it for what you want to happen. You are in complete control and in charge.

It is to your benefit to wear the crystal continuously (except when you go to bed and sleep; then it is not necessary to wear it). While you are awake, wear the crystal next to the heart chakra so it keeps you on course for what you really want to do.

Laying On of Hands and Love

A very effective method of healing is done by the laying on of hands and is one of the best holistic ways of healing that can be recommended. This has been practiced by healers since the beginning of time. Most everyone has experienced this kind of healing from a mother who has gently and lovingly laid a hand on a fevered brow.

The laying on of hands predates Biblical times, because when it was written about in the Bible, it had already been in use for thousands of years.

Laying on of hands can also be called magnetic healing. It is used by holistic healers to use one's hands to accelerate healing of the body cells. It is the most common method of healing and has been used throughout the world in every culture. The healer uses the magnetism within his/her own nervous system and concentrates on this energy to intensify its healing qualities. It is then transferred through the palms or fingertips (where there are chakras or energy points) to the patient.

The healer's hands are placed over the congested areas, actually making physical contact; or else are held a couple of inches above the body. Another method is for the healer to touch the head of the patient with the hands, while visualizing and sending the magnetism to the needed area.

Healing can occur gradually, or sometimes instantaneously. Patients have reported feeling heat, cold, prickles, energy surges or sometimes nothing. A biochemical reaction is thought to take place in the body chemistry of the patient from laying on of hands.

It is believed that this type of healing occurs in the spiritual body or subtle body and then interacts with and is copied by the physical body.

The laying on of hands is a very successful method of healing, too. For example, in the hospital when a nurse picks up a baby, the baby's body cells immediately have a *communication* with the cells of the nurse. The nurse has one focus of intent at the time, and that is to comfort that baby, so there is a communication. At one level of consciousness, the baby responds, "Gee, that's great. I'm being taken care of." This type of healing with physical touch and positive intent is tremendously effective.

Nurses are often unsung heroines for their healing efforts. Many nurses are angels of healing in disguise. They have this special touch and automatically know how to use it when they care for patients. Much of healing results from the nurse on the midnight shift who gently touched a patient or held them in her arms and assured them that they would be okay. When a patient is reassured with love and caring, this transmits healing to their body cells.

Newborn babies that receive no hands-on loving care do not thrive or live. Experiments similar to this have been done on monkeys. If newborn monkeys don't have a surrogate mother to hold and cling to, they wither up and die (even a stuffed toy has been used and is better than nothing). This need for a loving touch is part of the nature of all mammals.

It comes down to the communication system that transcends the five senses, because a newborn baby certainly doesn't think with his logical, rational mind. That hasn't yet come under control or been developed in this lifetime. What does happen is a direct communication through the cell level.

It's incredible to think about the intelligence that each cell possesses, and most of the time we don't credit our body with *any* intelligence—we think it's just the mind or brain that contains all intelligence. Yet the mind is just a small part of the wisdom our body contains. We fail to recognize/respect/pay attention to the consciousness the human body has outside of the control of the mind.

Now each cell certainly isn't a philosopher, but each knows what its purpose is, and they have this communication system so that they can send out danger signals, or the peaceful "all is calm" kind of signals.

It is important to send the body cells loving thoughts filled with gratitude for all the work the body does each day. The best way to do this is to use the crystal's amplification to relay the message to the body, loud and clear, that you love it and thank it for carrying you so well throughout the day. Look in the mirror each morning and tell the image you see, "I love you." Do this daily, even if you feel embarrassed or foolish, and I guarantee there will come a difference in the quality of your life and a marked improvement in the well-being of your physical body. Again, holding or wearing a crystal while you project love and forgiveness to yourself will amplify the benefits.

Love has a tremendous healing effect among all people. If you use the crystal along with feelings of love, well-being and devotion, that is the best healer of all. Love is the sense of caring, the sense of being benevolent. Used with a crystal to amplify its energy, love is a powerful healing energy.

Healing with Prayer

Prayers do work. The secret of a prayer's success is that the person who is praying has to believe in it. Use of a crystal greatly amplifies the power of a prayer. There is a difference between prayer and meditation. Prayer is talking to God, and meditation is listening for the answer.

If a person prays out of duty or projects doubtful thoughts such as, "I'm going to pray for Uncle George. He's an old geezer, but I'll go to church and pray anyway. It may work, but it's up to God. It's God's fault he's sick." It is unlikely the prayer will have any noticeable effect. These kind of thoughts show a lack of faith.

One of the most effective methods of prayer is to have a group healing. A prayer meeting where you have two people is about four times as powerful as one person praying. If you have four people praying, it is 16 times as powerful as having two.

It is helpful to have a small ceremony or ritual to build and focus the energy for prayer and healing. A simple ceremony such as having some lighted candles and a cup or bowl of blessed water will do. The water is symbolic of all eternal life. All life came from water, and all life is sustained by water. Water is an excellent medium for sending messages and for receiving answers. All present should be holding or wearing a quartz crystal for amplification of healing and the transference of energy.

The group present should then have a brief prayer service where they actually pray for whoever needs the healing. They should visualize this person well—in perfection, shining, beautiful, happy, and with a smile on their face. Visualize them the way they should be, the way they will be. Do not see them ill, damaged in an accident, etc.

Send this thought form and energy out. There is a special means of communication which has been set in motion. The prayer can reach thousands of miles—distance has no meaning—to the body cells of the person in need. The person receives the healing thought, the healing impression, and is helped. The person may not be aware that a healing prayer has been sent out, but they will feel better. They feel like something has happened during the night which has improved their condition. It is amazing that the prayer will reach exactly that person for whom it is intended. Maybe he/she lives in a city of hundreds of thousands of people, but that healing energy will come directly to him/her.

The group sending the healing and saying the prayer should visualize the absent person surrounded in a healing white Light and see him/her in perfect health, perfect body and perfect spirit.

It is helpful to have someone sit in a healing chair to represent the absent person being healed. All other healers and helpers stand around the healing chair and the volunteer. The primary healer stands in back of the chair and gives the volunteer representing the absent person to be healed an affirmation on a card such as this:

- Be calm
- Be at peace
- Put your trust in God
- Be serene

It is beneficial to have soft, soothing background music playing. This keeps everyone in a relaxed, alpha state which is conducive for sending telepathic messages.

As the person in the chair represents the person(s) who need healing, the healers can pray for all others on their list who need healing. They can softly speak aloud the names in order to channel the healing energy. The primary healer can hold his/her hands over the head of the volunteer to strengthen their power for the mental images to be projected. See those who will receive healing in joyful, shining perfection.

The healer standing in back of the volunteer now softly speaks a prayer for healing and gives the person a blessing. A short prayer is all that is necessary, from 45 seconds to a minute. If the healer has blessed or holy water, this can be sprinkled on the volunteer. Touch them on the shoulder, thank them, and they are released. This is a very effective way of healing.

Sleep Crystals

Very little attention has been given to sleep healing, yet it is a very vital part of the healing process. The best healing for the physical body occurs during the sleep cycle. Many fine physicians believe that 70 to 90 per cent of healing can take place during sleep because the conscious mind does not then interfere with the process.

The best time to program your crystal to heal you during the sleep cycle is to do it right as you are drifting off to sleep. Talk to the crystal when you are in the state between sleeping and waking (this is called the hypnogagic state). Programming done during this time will make the greatest impact on the subconscious.

Place your crystal on your left wrist, the same as for a dream crystal (see Chapter 7 for more information on dream crystals). The subconscious makes contact with the crystal best when it is next to your skin touching it.

Sleep crystals are controlled by your mind. The crystal amplifies what you program into it. You can program the crystal to revitalize and heal you. The important factor is your own belief. You must reprogram your subconscious. You must get through to it, otherwise the subconscious will keep on maintaining the previous programming. You must want to get well. You need to get through the five senses to the sub-conscious mind which knows how to change the reality. Your body and subconscious love attention, and will respond favorably to any kind notice taken of it.

There are many ways to use quartz crystal for healing, and this chapter just opens up some of the possible methods you can use. Crystals are remarkable enhancers of your own energies. Learning to redirect these energies for healing of yourself and others is easy and enjoyable once you know you can do it.

Photo by Frank Dorland

Electronic Quartz Crystal Handworking Pieces

Carved, Polished Quartz Crystal Handworking Pieces

Photo by Frank Dorland

Chapter Seven

PRACTICAL APPLICATIONS OF CRYSTAL USE

The use of quartz crystal can significantly improve the quality of each person's life. By programming a crystal for a specific goal, you can remove obstacles and problems which prevent you from being a satisfied, growing individual. A crystal can help you in just about any facet of your life, if you just stop a moment and think about how you can best utilize its help. Its assistance is always there, it doesn't demand any wages or payback, nor does it ask anything of you. It has no moving parts and will never wear out. It is harder and more durable than steel. Where can you find a better helper?

This chapter will examine some of the more practical and unusual ways in which a crystal can be used to make your life better.

Finding Lost Items with a Crystal

Holding a crystal in your left hand and talking to it will help you to find lost items. The crystal will amplify your sensitivity to the vibrations of the lost item, which is what you are trying to tune into. These vibrations will act as a beacon or homing device to help you locate what you are looking for. Ask the crystal to assist you in locating what you are looking for.

Self-Improvement with Crystals

There are many aspects of our lives which we would like to improve upon, but often we just can't muster up the will power. Again, the crystal can come to your aid. You can program a crystal to help you with any health issues such as losing weight, stopping smoking or changing your habits. For example, you want to quit smoking but just can't. You know you should stop for health reasons, but you don't have the will power, and the habit of smoking has been ingrained in your system too long (or so you think). Try programming a crystal and tell it something like, "I have quit smoking. I have completely lost my desire for cigarettes." Wear that crystal around your neck or keep it in a pocket, and whenever you want a cigarette, touch your crystal and feel the urge to light up dissipate.

Crystals Instead of Drugs

We are always looking for ways to expand our consciousness, because it brings us pleasure and heightened awareness. A shortcut (but a dangerous and unhealthy way) to achieve this is through drugs—but drugs aren't the *only* avenue.

From personal experience, I can attest to the fact that crystals can and do alter one's state of consciousness.

Wearing a crystal or having one close to your body helps you achieve a state of heightened awareness—a pleasurable state of being. Crystals have been known to stimulate or induce a state of euphoria. One explanation of how they bring this about is due to their consistent, undeviating frequency of vibration which can help to calm and regulate your body's vibrations.

And yes, if you are depressed, they could amplify that also. The good part is that you don't need to hang on to a feeling of depression. Simply hold your crystal in your left hand and speak to the crystal, programming it with something like, "I am relaxed, happy and carefree."

Crystals for Success in Business

Many business people are beginning to tune into their subconscious to help them in making more astute business deals. Many have a carved quartz crystal paperweight on their desk. Before making a deal, giving an order or making a decision, they will stop and put their hands on the crystal. The contact will heighten their feelings of intuition, broadcasting the correct answer as to what they should do. Sometimes they will go against their advisors, computer printouts, or the financial records, because they had an inner feeling to which they paid attention, and which was amplified to them by having a crystal near-by. By playing their hunch, based on what they intuited with the crystal's help, they have saved hundreds of thousands of dollars, or made a like amount.

The former vice president of Llewellyn Publications, Steve Bucher, has some interesting stories to relate as to how crystals have helped him recently. He attributes sales at a Northern California bookshow to the use of a crystal wand he had at the show. (For a more complete description of this wand and how to build it, see Chapter 8.) After a half day had elapsed, there were no sales—just small talk among the sales people—and the show was going nowhere. He was inspired to start waving his crystal wand, programming it to bring some sales. The others were laughing at him, asking Steve what he was doing. Steve replied that he had a new sales wand. Within ten minutes the first order arrived for Llewellyn. He then had to broadcast his success by again waving his wand, gaily saying, "It's working!"

Within a half hour two more orders were taken. Another major publisher quietly came up to Steve and asked to borrow the crystal wand. Within ten minutes he too had an order. All he could say was, "It's hard to believe."

Another beneficial use of a crystal pendant came about when Steve was returning from a London book show. After three minutes in the air, the airplane lost an engine. It was a precarious situation. The pilot decided to jettison fuel so they would have less weight and hopefully be able to return to Gatwick.

Steve put his hand on his crystal and prayed that if he could get on the ground again, he wouldn't complain about standing in line all day, rebooking a flight home. His plea worked, and the plane made it back to the runway.

The mishap created more problems when he was back in London, in line with 350 people and their luggage, all trying to reschedule a flight home. They were all going to a small, six-foot counter. While making small talk with another publisher, a thought entered Steve's head: "I have a crystal."

He put his hand on his pendant and silently spoke to it: "Okay crystal, if indeed things happen with you, it would be a good time to work your magic." Within four or five minutes a Northwest Airlines employee came near the group and asked some people if they were going to Minneapolis. Quietly Steve interrupted the employee and said he was going to Minneapolis, too. Within four minutes the employee came back, handing him a boarding pass, and instructed him to go to the Delta counter. Steve thanked his crystal and proceeded to go home.

Use of Crystals for World Peace and Healing

It would be very helpful to the Earth if people would meditate for world healing and world peace using a crystal. The more people who do this, the more energy that will be broadcast into the universal consciousness.

Using a crystal held in the hands or worn around your neck while meditating on this serves to broadcast the energies out so the healing and peace vibrations are felt in all countries.

The Earth itself is okay. It's just what humankind has done to it that needs correction. What we need to do is meditate and send peaceful thoughts to the governments so the leaders will focus their efforts on peace and positive means of leadership. This will help improve their mental attitude so that it's not one of war and mistrust.

Rituals and Crystals

Using ritual is a way of concentrating your energies on any given symbol of your choice in order to enter a different space—to produce a shift in consciousness which will allow you to align your psyche or inner being with those forces which the symbol typifies. The ritual is a tool which helps put you in touch with your inner self.

In ancient times, a ritual was a natural means of accessing the power within that each of us has. There are many different types of rituals —religious rituals, magic rituals, business rituals, school rituals, etc. Using or wearing a crystal can greatly enhance the power of a ritual.

It has become common practice to include a

crystal in many magical rituals. One magician says that to him a crystal represents the element of the Earth. A friend uses a crystal on an altar, facing North.

Sex and Crystals

A crystal is helpful for erasing old, negative programming regarding sex. Sex can be a beautiful, wonderful experience, depending on what is in your mind. Basically, the mental image or thought of love and creation is what good sex is all about.

One woman reported that wearing a crystal on a cord around her hips greatly increased her awareness of sex and her sexuality. Her husband was very pleased with the effect the crystal had on her!

Healing Pets and Crystals

A programmed crystal can be very effective in healing or helping pets. One woman reported that she effected a major change in the health of her dog by using a crystal.

She had a well-loved dog, who grew old and manifested arthritis. He was so crippled that he couldn't get up the stairs to the second floor in their home. That was his favorite place to spend the day, because that was where the sunshine was, and he liked to sleep in the warm sun. The lady would carry the dog up so he could enjoy his favorite spot; but later he would whimper and whine because he couldn't come down the stairs. So she would have to make another trip up the stairs and carry him back down.

She then got the idea to knit a small bag for a crystal so the dog could wear it around his neck. Before she placed the crystal in the bag, she programmed it. She put the dog in her lap, and then held the crystal in her hands and talked to it as if she were talking to the dog. She told him how he was going to get well, and that she was fixing him the best food for his well-being, and that he really was going to be well.

In about ten days the dog was running up and down the stairs. She was close to her dog, and she programmed the crystal so it would respond to the dog's vibrations. The dog sensed that somebody really cared about him, that something was happening for his benefit. In some mysterious manner this triggered his immune system so that it rebalanced the body chemistry. Whatever was causing the arthritic pain disappeared so that the dog could function again and walk with comfort.

After she programmed the crystal and put it around the dog's neck, the message was being continuously fed into the body cells of the dog. He was constantly receiving his mental therapy, which was activated by the nearness of the crystal to his body. This technique was effective because the owner believed it would work, and because of the love and communication between her and the dog. It was not necessary for the dog to think. His body cells responded to the trust, faith and communication between his mistress and himself.

On a lesser note, I have cleaned and programmed crystals for health and well-being, and put them in goldfish bowls. The fish are certainly active and lively, and they seem healthier.

Sound and Crystals

All matter is composed of vibration, and how we identify different colors, objects, etc. depends on their rate of vibration.

There is an excellent book by Elizabeth Keyes called *Toning*. In it she explains an ancient method of healing called, appropriately, toning. It has to do with sound and vibrations, and the mechanics of letting the voice express itself.

Everyone knows the value of groaning when trying to do hard work—the body just naturally lets the sound out. When a woman screams, she is releasing a huge amount of tension that would otherwise be stored in the body in a negative manner. Similarly, laughing at a good joke is eminently healing. Laughter is said to be good for digestion, too.

Look at what Norman Cousins did with laughter and self-healing. As you may recall, he wrote a book describing how he recovered from a debilitating spinal disorder with the help of humor, by reading books that made him laugh, and watching funny movies, while he was recuperating.

Use of a crystal with certain kinds of toning,

chanting, or singing, or just "letting go" with the vocal chords, is very beneficial to health. Quartz crystal is a tool to aid in healing with sound.

ELF Wave Protection with Crystals

It is important to understand the human body and its electronic network. The human body is probably the most sensitive receiving apparatus the world has ever seen.

The body is 75-80% water, and water is one of the basic conductors for electronic (electrical) use. The body is an efficient receiver, like an antenna.

We are all open to the reception of different wavelengths. In fact, we are submerged in a sea of radioactivity—we are constantly being bombarded with wavelengths of all descriptions: high, medium, low and all varieties. Most of us don't realize that this bombardment exists.

The constant bombardment of deleterious wavelengths in some of the larger cities has been blamed for some kinds of insanity and nervous disorders. At any rate, we need to protect ourselves from those wavelengths which have a harmful effect on the body.

It is commonly believed that ELF (extremely low frequency) waves disrupt and hinder the body's immune system responses. A common source of ELF radiation comes from high power lines (and some antennas used by navy submarines.)

To use one city as an example, in San

Francisco people are bombarded with Coast Guard and radio broadcasts, the weather control tower, and the RCA overseas radio communication center. That complex rebroadcasts information for the entire United States. Then there are the radar towers on the different mountains in the area, to say nothing of the ships at sea broadcasting information. ELF waves will pass through water and most other barriers.

Some researchers feel that ELF waves may be a factor in dolphin deaths. They believe that the dolphins' exposure to ELF radiation (especially from submarines) was a primary factor in their dying. The dolphins in question appeared to have been severely stressed, and died of bacteriological infections.

Every ambulance, every fire engine is doing the same thing—emitting wavelengths, audible or inaudible, that may not be conducive to good health. Our homes have color TVs and microwaves which give off radiation. We are living submerged in a sea of radioactive waves all the time.

Sometimes these vibrations are intelligible, such as in a telephone conversation or while tuning in a radio. The rest of the wavelengths we don't consciously understand, but they do exist, and they do affect us.

Our bodies are receiving them, and our body cells are reacting to them even though our five senses are not aware—they are blocked. These vibrations are deleterious to our health because we're constantly in a turmoil without even being aware of why we're being stimulated.

Again, use of a crystal can help protect us from the effects of these wavelengths. A crystal programmed to let only beneficial vibrations reach our understanding is helpful to wear or carry in the pocket at all times. A crystal may be programmed to welcome all beneficial vibrations and exclude or discard every negative or harmful radioactive signal.

Famous People and Crystals

One of America's most famous and best-liked inventors was Thomas Edison, who lived from 1847-1931. Although best known for his work with electricity, Edison was very interested in life after death and psychic phenomena. It is said that he slept only four hours a night—thus adding ten more years to the conscious part of his life.

He was fascinated with quartz crystals, and mentions them in his diary: "Even in the formation of crystals we see a definite ordered plan at work."

Edison used crystals for creative thinking. He would visit a ranch for short vacations. In back of the ranch, up in the hills, quartz crystal deposits could be found. Edison would hike up to the hills and dig crystals out of the pockets. He enjoyed flashing them in the sunlight, seeing the way the Sun reflected on them. Then he would put them in his pants pocket and walk back down to the ranch house for lunch. It was said that he was an uncommonly big eater—he'd eat a lunch big enough for several people. Next he would come out to the front porch and sit back, just lolling around with his hands in his pockets, fingering the crystals he had found, grinning and looking at the sky. He would see all kinds of images in the clouds, and enter a relaxed state of reverie. These were his "dream crystals," and perhaps they helped him tune into the right, intuitive side of the brain.

An inventor usually will first visualize the things he or she creates, and then make what they have seen. Many inventors have said something to the effect of, "I saw it. I don't know who invented it, but they showed it to me and I saw it, and then built what I had seen."

The same thing is true for many composers who say that wearing a crystal helps them to hear the music they create. Several say that it is helpful for them to meditate while wearing a crystal. When they come out of meditation, they write down what they've heard.

If you too want to use this technique to create or compose, take with you into your meditation an imaginary blackboard. When you see or hear inspired notes or words, write them down on your imaginary blackboard. By doing this, you bring into action a different part of your brain than you would have by just seeing it. You compose your information into alphabetical letters (words), and by so doing you go from your right brain to your left brain to write it down in the meditation. Then, when you come out across the threshold into normal, waking consciousness, you can better remember what you heard or saw in meditation. Otherwise, the things you were aware of in the dream or alpha state are likely to dissipate very rapidly when you return to normal, waking consciousness. Using an imaginary blackboard or writing tablet is a big help in retaining what you saw or heard in the alpha/theta state.

Another famous person who realized the value of quartz crystal while still very young was the English poet and writer William Butler Yeats (1865-1939). As a child he was very interested in occult lore. The visit of a young Brahmin man from India left a very profound effect on the young Yeats and some of his friends. Yeats reports on his experiments with quartz: "When we were schoolboys we used to discuss whatever we could find to read of mystical philosophy and to pass crystals over each other's hands and eyes and to fancy that we could feel a breath flowing from them as people did in a certain German book."

George Sand (1804-1876), the French novelist who was famous not only for her romantic novels but also for her numerous love affairs, used a crystal. As a child, she discovered an ability to see pictures in a crystal ball. In later years, when writing, she formed a habit of concentrating on the crystal for inspiration.

Another very famous person, Alice Bailey (1880-1949), Theosophist and author, speaks about crystals in her famous book *Esoteric Psychology:* "We have in the mineral world the Divine Plan hidden in the geometry of a crystal. If you could really understand the history of a crystal, you would enter into the glory of God."

Electricity and the Body

Not only does the human body operate on electricity, but also the whole universe runs on electricity. When dealing with crystals and healing, it is important to remember that crystals are electronic.

We don't know what the source of the universe's and the body's electricity is. There are some things we don't understand about electricity. It is an elusive phenomenon. Scientists do not really know what electricity is. They know what electricity causes and how harnessing it creates energy, but *no one really knows what it is!*

There is a unique kind of electricity coming out of the human aura. It is composed of different kinds of electrical vibrations. Our body magnetism is a combination of a collection of direct current and alternating current (it does have a positive and negative polarity), and some more unnamed glues which hold these electrical energies together. This is an amalgamation of the secret power—and that, possibly, is what God is.

If we could theoretically remove the positive and negative of electronics and magnetism, the universe would be in chaos. Everything would go to pieces and there would no longer be any coherent order.

The human body runs on positive and negative electricity. Each cell has a positive and a negative side. The whole nerve network is nothing but electricity. The value of a crystal is that when you pick it up, it starts vibrating on an electrical frequency that is in harmony with your body and your subconscious mind.

The crystal can be likened to a choir that comes in, not on the same note that you are on, but on these various harmonizing notes which boost and simplify. Instead of one little weak voice of your subconscious which you can't hear because of the TV, the telephone ringing, etc., you have a whole choir swelling up with these feelings, these electrical impulses. The "choir" boosts information out of your subconscious, things you know but are not yet in communication with.

This amplification of the harmony of the "choir" brings things—your hunches, your perspective—into realization so suddenly that you have a mind-blowing certainty that you've arrived somewhere. You don't exactly know where, because you need to manifest this awareness more, but you are now better able to make an intelligent judgement as to what is happening in your life.

The crystal will boost your understanding to help you realize that here on Earth you make your own heaven or your own hell. You are creating your reality. You are either destroying the body or helping it to work properly. This is what the crystal is good for—to help you realize the good things you can accomplish and become the beautiful person you were meant to be.

Crystal Radio Sets

Many proponents of quartz crystal like to state that early crystal radio sets operated on quartz crystal. This is not true.

Crystal sets were actually small radio receivers consisting of a tuning coil and a crystal (not quartz) detector stage to demodulate the received signal. They had no amplifier.

The crystal which was used was galena, an isometric form of lead sulfide. It is blue-gray in color, and is a brittle substance. Galena is found worldwide, and in the United States there are extensive deposits near Joplin, Missouri. Common sizes range from one-half inch to two inches with occasional specimens up to 24 inches. When heated, galena emits sulphur fumes, then melts to a lead glob.

In a crystal radio set, the one-half inch size cubes were used. Electronic contact was made with the crystal with a skinny, flexible "tickler" wire called a cat's whisker. By moving the cat's whisker around over the crystal, a sensitive area could be found, and then the radio could be turned on.

The older, earlier types of earphones did use quartz crystal in their mechanism, however.

Experiment Showing Crystal Light Phenomenon

There is a simple and interesting experiment to conduct which shows the ability of crystal to generate light.

Take two pieces of naturally formed crystals (the larger the better—hand-held size is best). Make sure that each has a flat, unpolished (sandpaper-like) side. Take these into a darkened room like a bathroom or closet where there is very little or no light. Briskly rub the two flat sides of the crystal together. Amazingly, they will produce light, almost like a fluorescent tube that is flickering and warming up.

As long as you rub the two flat pieces of crystal together, producing friction, you will generate light. No other rock, gem or stone will do this. Only quartz crystal has this unique ability. This is a demonstration of its piezoelectrical quality—the pressure from squeezing and releasing quartz that generates electricity.

Dream Crystals

You spend approximately one-third of your entire life sleeping. Some regard this as wasted or lost time because the conscious mind has no recollection of what occurs during much of this period. With a dream crystal you can now tap into this facet of your life and use the information found in your dreams to make your life better. Dream crystals are fascinating tools.

People who have suffered from recurrent nightmares find that going to sleep is not a pleasant time. A repeated nightmare that a person can't seem to escape from is a horrible experience. It has been discovered that if you program a clear, polished, smoothed quartz crystal to change or improve your dreams, you will definitely note an improvement in the quality/type of dreams, plus you will have better recall.*

In order to do this, secure it with tape to your left wrist so it is close to your body, especially the pulse area. Make sure that the crystal is clear, and that it has no sharp edges which could cause discomfort while sleeping. Program the crystal, telling it that your subconscious programming is faulty, and the memory of this horrible experience is only a dream. It is not a beneficial thing, and you choose to stop having this kind of a dream. If your subconscious ever does repeat this nightmare, have a bell that will ring and a voice that will say, "This is only a dream. Wake up." You can banish a bad dream. If there is a horrible scene taking place, you can automatically build a brick wall to protect yourself. If there are ghoulish figures, these figures can be evaporated into steam and disappear in a cloud of vapor, be dissolved in a white light, or sent up into the Sun to be burned or absorbed. You can program your dream crystal to prevent you from ever having a nightmare again.

*Dream Crystal Kits are available from Llewellyn Publications.

If your child has nightmares, you can hold him/her in your lap and lovingly tell them that a dream crystal can help them sleep better, too.

Many people will program a crystal for what they want to dream. They can actually program their dreams. It is also helpful to program your crystal in order to better recall your dreams upon waking. To replace bad dreams with good dreams is a function that the dream crystal can help you with, too. This has worked with many people. Some people put a crystal under their pillow, but it has been shown that a crystal needs to be close enough to your body to touch it, within approximately one or two inches. If you put it under your pillow, it generally won't be close enough to be within your etheric field. If you say, "I have placed a crystal under my pillow or mattress and it now is going to work," it, of course, would have some effect. The best method is to have the crystal within one or two inches from your body to receive the feedback electronically directly into your body cells, which in turn go into your subconscious memory bank which is responsible for stirring up these dreams. It is only in your mind—these are only subconscious memories lurking from somewhere.

It is known that when you are born, you already have thoughts and memories in your mind of things that have happened, as babies still in the womb are already dreaming of events. How can they dream of events when they haven't been born yet? Their eyes have never been open, yet they have this memory of something. Where is it from? Is it a past life? Or is the child tapping into the memory bank of the mother? Is the child reliving experiences of his ancestors? The moment babies are born, down falls the curtain of forgetfulness, and all of their memories are washed out of their five-sense mind, in part because the doctor slaps them on the bottom, the nurses are cuddling them and bright lights surround them. It is interesting to speculate on what babies still in the womb can possibly be dreaming/thinking about.

If you dream more and remember your dreams better, it is likely that you will need less sleep time. An additional benefit is that you will be less tired during the day, when you are in the regular, conscious world. Part of the necessity for sleep time is to sort through and work out the day's problems during the night in imaginary scenarios. This acting out of possible scenarios frees the person from actually having to go through with certain experiences, because he/she already knows the possible outcomes and what the learning will be. With the dream crystal programmed to help you remember your dreams, your sleep will be more productive.

Do not worry if you fail to get an answer or can't seem to remember your dreams immediately upon waking. Sometimes it will come to you later in the day, or even several days afterwards. This happened to me when I first started using a dream crystal.

I had programmed (told) my dream crystal to give me an answer to a health problem through a dream. When I awoke the next two mornings, there seemed to be no answer. I was disappointed and thought, "These dream crystals don't work. They're worthless." Then on the second morning, as I was folding laundry, the answer came as clear as a bell; I knew what I had to do to improve this particular health situation.

So take heart. Sometimes the answer to your queries may be delayed. Another thing to be aware of when first using a dream crystal is that you *may* have bizarre dreams. This too happened to me. I had four or five strange and seemingly chaotic dreams. Perhaps I was just doing some mental housecleaning, removing the cobwebs from my mind; or maybe I dreamed like that all the time, but just never remembered such crazy dreams!

Experiencing Past Lives with a Crystal

At no time in history has there been such universal fascination with and deep interest in reincarnation, our lengthy sequence of past lives.

Most people are fascinated by the concept of reincarnation and have probably wondered if they have indeed lived before. The only way to convince yourself whether or not this is true is to experience a past life yourself. Then you can be sure for yourself. If someone else tells you about a past life it is not so believable as when you relive one yourself. The crystal, of course, is your key to help you unlock this fascinating door.

There are several different explanations for past lives. One could be that when you tie into the subconscious memory bank, you relive and share the experiences a long departed relative had—perhaps a great, great grandparent. It is believed that with the right suggestion we can tune into the life of any person who has ever lived.

One explanation for this is that a "sensitive" can read the akashic records to obtain this information. The akashic records can be likened to a huge aura surrounding the Earth wherein are kept all memories, all experiences, all vibrations. This giant aura serves as a huge, complete library or computer where all records are kept of everything that has ever happened in this world, as well as all the thoughts anyone has ever entertained.

The other, more standard explanation comes from the concept that you as a soul have chosen to incarnate in many, many lifetimes to learn certain lessons in order to become the god being that you are capable to becoming, and to understand that you were a spark of the Divine when you were first created.

If we complete our required lessons in this lifetime and pass our examinations in schoolroom Earth, we will advance to the next step of growth and evolution. We learn what we have set out to learn, and then we can take our experiences with us for a better life next time and won't have to repeat the same grade. Each life can be better and more enjoyable.

If this theory is true, our subconscious memory bank has stored all the experiences that have ever happened to us. These cell experiences of the more dramatic things that happened in our ancestry, which go back maybe three hundred thousand years, are still with us. None of these things are ever forgotten, and that is why we act the way we do. That is why we have certain characteristics, certain actions and certain reactions.

A crystal can help us tune into that information. If a crystal is used in meditation to its highest advantage, it will provide information firsthand about other lives you have lived.

Here is one technique to help you access a past life, or a past-life experience: Before the meditation, form a question about a life or experience about which you want information. Write down this question very clearly so that it is concise, logical and understandable.

You begin your meditation and remove yourself from your five senses. Next, you gently concentrate on the question: where you came from, what are the lessons that you need to learn, etc. Keep it simple—one question at a time.

Then you enter into a more relaxed state with your crystal. The crystal should either be around your neck close to the heart and solar plexus area, or held in your hand.

By doing this it will be possible to receive specific answers to questions you have written down. In this way, you have formulated the question in the five-sense, rational, logical mind, and transferred it to the subconscious memory bank. The subconscious will then unlock the information from the cells, where it is stored.

The crystal will amplify these signals, which are electronic. This can be likened to receiving a radio broadcast from your soul, but the volume is turned so low that you are not hearing the voices. You are able to hear just little murmurs and soundwaves without any understanding or meaning. By using a crystal, you are able to turn up the volume and tune in the vision or remembrance more sharply. Suddenly, you will have the visions and the scenes and you will swear that you are there and actually living the situation out. You will be watching it and observing everything that is occurring, and it will be so lucid, so clear. People who have been able to tune into past lives in this manner have said, "I was there, and no one can tell me I wasn't there. I went through that and really experienced that."

This can be a mind-blowing experience because you may find yourself in a life-or-death situation. You are either going to perish or else you are going to escape from the situation. No question about it, it is crucial at the time. There will be clarity that is more vivid and real than many things you have experienced in this lifetime. In this physical life, when a crisis appears, often everything is confused and unclear that we are unsure of what is going on, and we often simply react. In the meditations, everything becomes so clear and so vivid that you not only know what is going on, but you are there taking part in it and seeing yourself at the same time. That is you, and you are there.

Generally, you will not experience these kinds of traumatic lives until you can handle them. That is why past lives are almost always blocked from the conscious memory. We oftentimes could not deal with the things we have done to others and what has been done to us. It is very unnerving to go through an experience of being tortured or being hunted by someone. (We all have done just about everything. We have been both murderers and the murdered. We have experienced both sides of good and evil. We have been both cruel and kind.)

I had an interesting past life recall of having my right hand cut off with a sword while I was lying on a cobblestone street. I had the feeling it was in Rome, during the days of chariots. I was wearing a soldier's uniform. It was a horrible experience, true, but until I relived that scene, in this lifetime I had always been terrified of getting my right hand cut off or sawed off. When I had this past life experience, I was freed from any more fear about my right hand being in danger in this lifetime. I then realized the hand accident had already happened in the past and it wasn't something I needed to worry about any longer.

Using the crystal to experience past lifetimes and traumatic events helps us heal so we don't have to keep these blocks—it helps us heal old "wounds," as it were. It removes blocks to our further development.

These life experiences are the foundation we've built over eons of time, and we don't have to repeat the same lessons over and over once we become aware of what we have actually learned. Each experience has been added to our foundation so we are much better off now than before. Adding these experiences to all of our lives can be compared to grade levels in school. A ninth grader knows much more than does a third grader, and is able to draw on more past experiences and learnings to make a decision.

Crystal Learning

The quartz crystal can be a wonderful help in assisting you to learn new information. The crystal is able to process information holographically, as compared to the computer, which stores and retrieves information bit by bit.

Ancient people, especially in Atlantis, used the crystal for storing information. It stores and retrieves knowledge as whole, intact images which don't need to be converted into parts, as they do with an electronic computer. Ancient men of science, and seers, could mentally project their knowledge, whole and complete, into the natural micro-computer field of a crystal. Whenever they needed to retrieve their information, they attuned themselves to the crystal, and stored images were reconverted into the original concepts.

In today's world of learning, a mind/crystal interface is an interesting possibility. A student could learn math, art, science, etc., with their "crystal computer." The crystal computer will have a program of stored-image subjects. When the student tunes into the computer and becomes "at one" with it, they can easily process huge amounts of information both instantaneously and holographically. S/he then knows things from the inside, and the immediate learning task is to translate what is known with the right brain into the logical left brain, which processes knowledge verbally.

Future Use of Crystals

It is said that there are twelve crystalline fields of communication within the Earth. People of the future will be able to communicate with the distant galaxies, and may use crystals for this communication. These twelve crystalline fields will be discovered throughout the world in subterranean tunnels or channels where crystal energy has been previously used to outfit a scientific technology. This crystalline network is a system of channels which connect with resonating crystalline structures for image and information processing. There are grid mappings that are presently not understood but which will be used as coordination points for many fields of communication to overlap with the Living Light in the Universe. These twelve grids act as focal points for the transmission of faster-than-light particles. Crystal communication is thought to be capable of going beyond our electromagnetic spectrum by being activated by the proper psi grids aligned with the twelve crystalline chambers built into the grid structure of the Earth. Astronomers can then use this crystal latticework to bend light waves for universal communication.

A new word has been coined—"crystalogy." It refers to the study of crystals. Some New Age people feel that this will be one of the sciences of our future.

Crystals for Psychic/Spiritual Development

Quartz crystal is an excellent tool to use for clear channeling of your Higher Self and its wisdom, and help into your ordinary life. Recall that crystals are individual sources of perfect form, each having its own frequency of vibration for focusing your positive thought forms.

Crystals help us in becoming more complete, and in taking more responsibility for our actions. They guide our footsteps on the path to spiritual awakening. They lovingly help us build a greater harmony of mind and spirit to become the radiant beings that we all are.

Meditation and prayer create powerful rays of White Light, which is like food for the soul. This White Light stimulates the pineal gland (also known as the third eye). Holding a crystal in the left hand during meditation steps up the frequencies and energies for the pineal gland. A small crystal taped on the forehead over the third eye is a powerful charger for meditation.

Focused gazing or meditation on a quartz crystal can stimulate the pineal crystals, through the eye ray, bringing about the development of clairvoyance, clairaudience, and clairsentience. One method of developing clairaudience perception is to hold a crystal in your hand and combine its power with chanting. Each crystal holds within its structure a sacred code of sound peculiar to it alone. This sound can be discerned by chanting and focusing attention upon the crystal you have chosen.

It has been said that in the great Temples of Initiation, the secret password was encoded in quartz crystal. The person seeking initiation would have to hear the password clairvoyantly from a spirit guardian before they would be able to enter. The right vibration or sound would automatically open the portal, much as our automatic door openers work now by Light breaking a barrier.

Another way in which crystals help your spiritual progress is with their rainbow light. Attach a string to your crystal and hang it in the window so the sunlight shining through it will shower you with the colors of the rainbow. Sit in this prism of rainbow light if you can. Otherwise, visualize the light rays coming through your crystal. This light will stimulate your etheric body and increase your auric protection against adverse influences, and raise your level of consciousness.

Quartz crystals represent crystallized energies of the Sun and light frequencies. An average person is like a walking crystal antenna who is out of tune! You can fine-tune your body and bloodstream, much like harmonizing a song or tuning a violin, using the energies of crystals.

Myths About Crystals

Quartz crystal is the most marvelous tool there is to cultivate the human mind, because you actually get feedback from it. Some teachings would have you believe that it is not the human mind, but it rather the way that you place the crystal—whether it's upside down, or right side up, or hang it with the point down, etc. that causes its effects. This is not important. What *is* important is that you learn to use your own mind for focusing your intent. You have control over what you think and what you do. The crystal will amplify what you tell it. The theories which maintain that the crystal has power only serve to diminish the person's own power. This concept keeps people from advancing to the point where they will think for themselves and take control of their own lives. It is like giving away one's power to depend on some mechanical type of thing, i.e., a piece of mineral which really doesn't work by itself. It is necessary to have the mind interface with the crystal in order to set its energies in motion to change your life.

Many of these theories and stories are interesting and we would like to believe in them. Many of them take the powers of the human mind and put them into some imaginary process. Most of this is created in someone's fertile imagination who wants to teach classes and charge money, and so they believe that they must have something different, something that is saleable.

Chapter Eight

MAKING CRYSTAL TOOLS AND AMULETS

By now you may be wondering how you can learn to fashion your own crystal healing pieces. As of yet, the supply of rounded, smoothed and polished quartz crystal is not plentiful so this type of quartz is not readily available. However, it is fairly easy to make simple polished healing and working tools which are very effective.

It is helpful to know someone who has lapidary equipment (most communities have a rock or mineral club which will do work for hire). Otherwise, it is necessary to purchase some basic lapidary equipment—at least a very simple rock tumbler.

You can use a piece of natural quartz crystal which has the six flat sides and the six terminated sides which look like they are faceted and end in a point. First, put the crystal in a tumbler and knock off the sharp corners and the sharp points, because a good crystal that you use for healing or meditation should not have pointed edges. Finally, polish the crystal. By smoothing the crystal it becomes more responsive and is more effective for amplifying and broadcasting energy. The energy that is inherent within the crystal is allowed to come out faster, in a more efficient manner.

However, if you don't have any lapidary equipment, you can still obtain good results from a crystal by wrapping your fingers around a regularly formed crystal to break the "hall of mirrors" effect and stop the energy from ricocheting. By completely enclosing the crystal in your hand, you can achieve maximum contact so more energy goes into the body cells. Otherwise, some of the crystal energy will be wasted.

If you have lapidary equipment, grind away the imperfect parts, such as the bottom and all the areas that are cloudy. Cloudy areas, which are full of inclusions, indicate that those parts of the crystal are full of impurities. The impurities ground out the flow of electrical energy, and it is the energy that you want to enhance.

Next, shape the top so it is attractively formed for whatever you want to make out of it. For instance, biocrystallographer Frank Dorland will grind a crystal to suit an individual. He will pick up a crystal, hold it in his hand, turn it over and over to become acquainted with it and feel its energies, and then just sit down and saw off the bad parts. He then has a rough form to work

from. Next, he grinds away on a diamond wheel
the sharp corners, the sharp edges, the flat surfaces so that everything is rounded. He attempts
to form both concave and convex surfaces.

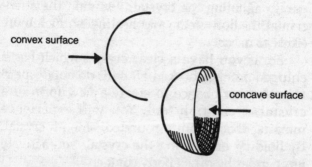

convex surface

concave surface

The energy will be broadcast much more
easily and in a greater area when using this
reverse curve. An analogy can be used with the
early radio broadcasting stations which used
one long, slender pole. At the top was a rounded
ball. This type of receiver antenna is still used on
automobiles. The round ball forms an area
where the waves can be broadcast out in a spiral
form, in a pulsing rate. The energy transmissions
are released more easily, without any impedi-
ments so the energy is not wasted.

To make a very modest crystal healing piece,
one end should be rather rounded and gentle, in
as large a portion as possible. The other end can
be more or less pointed, but not with a sharp
point. The gently rounded area could be used
for broad areas of the body, such as a muscular
area or for a headache, in order to broadcast the

energy beams into as wide an area as possible. The other, smaller and more pointed end would be used for acupressure. By pressing the pointed end gently against an acupressure point and gently agitating the crystal, you gain the attention of the body cells, and healing is much more likely to occur.

So, if you have a clear crystal which has a chipped point or broken-off part, do not despair! Here is your chance to create a new form of a crystal/divination tool. You will experience more feedback from a rounded, smooth crystal. By holding or wearing the crystal, you activate its energies by direct body contact.

It is not necessary to spend a great deal of money to buy an elaborately carved crystal to start out with, because it is your experience that will help you. A simple smoothed and polished crystal will provide a great deal of feedback, and is as effective as a much more expensive piece. The crystal already has the energy in it, and polishing and smoothing it makes more of that energy available.

Although the "voltage" from smoothed, polished clear quartz is high, some individuals will not experience any feedback from even these crystals. No matter what these people do, they are unable to sense anything from a crystal. Most of us are less sensitive when we first begin to work with crystals, but go on to develop a greater sensitivity, and in time discern a clearer feedback from them.

Photo by Frank Dorland

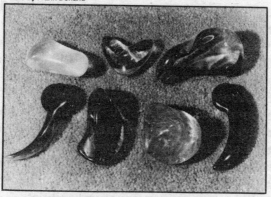

Quartz Crystal Healing Tools

Photo by Frank Dorland

Polished Quartz Crystal Handpieces

If you are making a pendant, the metal of choice to use for fastening it and for the chain is silver. A "purist" would only use silver, and they believe that only silver should touch the crystal. This is the classic and traditional belief. If silver is not available or within the budget, silk or cotton are compatible materials which can be used to fasten a crystal around the neck. Again, the chain or cloth should be long enough so that the crystal hangs low between the heart area and the solar plexus. These areas of the body contain the most power and energy, and the crystal worn here will best amplify your natural abilities.

A programmed quartz crystal worn on the body in this manner is the easiest and most efficient way of constantly supplying the body cells with helpful energy and positive programming.

Silver is important to the subconscious, and has a greater effect on it than on any other metal. Silver is related to the Moon and to water. The Moon speaks to our subconscious, and water sustains life (the human body is mostly water). Quartz crystal is solid, yet looks like frozen water. It has been called holy ice. Gold and the Sun deal with commerce, the day activities, while silver and the Moon deal with night, the subconscious, intuition, and mysticism.

Crystal Amulets and Talismans

Amulets or talismans have been in use since the earliest times and are always a source of fascination for humankind. Some of their reputed powers have only a placebo effect, for example with a rabbit's foot or a lucky coin, but some have actual magnetic or magical effects on their users. If these are carved out of quartz crystal, they have proven to be remarkably effective for those who use them.

An amulet, either natural or constructed, is believed to be endowed with special powers of protection or to be able to bring good fortune. Amulets are carried on the person or kept in the place where its influence is desired, e.g., a bedroom or on a roof. The word amulet comes from the Old Latin *amuleteum*, which translates as "means of defense." The general modern connotation of an amulet is something that is carried on the person or worn on the neck.

Amulets have been used (and still are used) by people of all ages and cultures. They are thought to derive their power from their connection with natural forces, from religious associations, or from being made in a ritual manner at a time of favorable planetary influence, from a suitable material and in a special shape befitting the purpose. Certain natural stones (especially quartz crystal) without markings or carvings have been used as amulets by many cultures.

A talisman can be defined as an object bearing a sign or engraved character, and thought to act as a charm to avert evil and bring good fortune. When it is marked with magical signs, it is believed to confer on its bearer supernatural powers. It too is generally placed near the

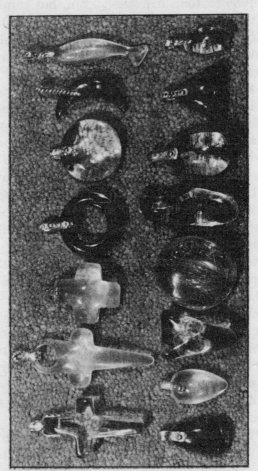

Photo by Frank Dorland

Carved Quartz Crystal Pendants

desired location of influence, such as on the body or in a place in the home.

Whatever you choose to call them—amulets or talismans—these lucky pieces made from quartz crystal will have an actual charge, and will benefit your life and protect you. They only need to be programmed for the intended purpose by speaking your intentions out loud to the crystal object of your choice.

Symbols in Quartz Crystals

There is an interesting symbol that resides in the structure of a naturally formed six-sided length of crystal.

Since the crystal is six-sided, it has two, three-sided triangles which make a Star of David, another very powerful, archetypal symbol. Connecting the outside of the Star of David—the straight lines from the points— makes the six-pointed star. The scientific classification of quartz crystal is trigonal, or the six-sided system.

The Star of David is a very ancient symbol which predates the Biblical David, and it goes back to the double trinity.

To carry the understanding of the two triangles further yields some interesting correspondences. Each of the triangles represents an aspect of the holy trinity. One of the trinity (triangle) is positive, and one is negative. Interestingly, this coincides with some of the scientific findings of the attributes of quartz crystal.

When a crystal talisman is cut from a positive to a negative face, it is more efficient.

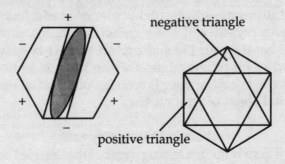

Six-Sided Cross Section of Quartz Crystal showing Star of David

An oscillator is cut from a corner to a flat surface, so it goes from one triangle through the center and to the side of the other triangle, thus giving the positive and negative on the edges.

The triangle is considered a perfect figure, signifying the completion of something. It represents spirit in three modes of expression: will, wisdom, and activity.

Most religions have the trinity represented in one way or another in their symbolism, and the number three is emphasized. The significance of three originally came from the father, mother and the child—the family, the completed aspect of the duality. It represents both positive and negative polarities, plus the result of their interaction, i.e., the completed family.

Triangle

The triangle is considered a perfect figure, signifying the completion of something. It represents spirit in three modes of expression: will, wisdom and activity.

Cross

The cross unites the material world with the spiritual world. The cross implies vital action and union, and manifestation of creative forces in material ways.

Circle

The circle signifies cosmic force or the cosmic mind. It is the beginning without end. It is a spiritual symbol of the realization of perfection, and release from bondage.

Ankh

The ankh represents the union of the male and the female. The dynamics of the two blended together produce a balance which allows life to sustain itself.

Common Archetypal Symbols

The use of the shape of a cross has powerful, archetypal connotations when carved from quartz crystal. Its significance can be traced back to many different cultures throughout the world from earliest times.

The cross is thought to have originated from what ancient man saw in the air on a cold day when worshipping the Sun. This symbol can be seen when there are ice crystals in the air, and a cross will appear to be in front of the Sun.

Because of this unique, mystical phenomenon, the cross was one of the first symbols of God, the almighty power in the sky.

The cross in all its varied forms (some sources say there were over 200 forms of the cross in use before Christianity began) evokes a powerful archetypal, universal message which has been programmed in everyone's subconscious memory bank.

Crystal Power

We know that crystals have captured the imagination of people since the beginning of time. The structures of crystals are aesthetically pleasing as well as precise, implying an order to the Universe which is reassuring to the pattern-seeking mind of man. But, crystals are more than beautiful examples of the order of the Universe. They are also amplifiers of the Universal forces and can be used to modify, accumulate and direct the powerful mental, psychic and physical energies we all have.*

---

*See *Crystal Power* by Michael G. Smith, Llewellyn Publications, 1985 for complete details on how to build these and other devices.

What you think is what you get. You're probably familiar with most of the positive-thinking procedures and how they apply to your life. They always work. The only difference is how fast or how slowly they work. With the use of crystal devices, you can learn how to cut down on time and effort. Your time is important! Also, the tools amplify your intent many times over.

One tool you can learn to make is an Atlantean Power Wand, and its most valuable use is for healing. In the process of healing, the energy from the wand is used to remove energy blockages in the biomagnetic field of the body. This provides a balanced form of the body's life-force energy. Removing the energy blocks provides a balance, while the body actually does the healing itself. This type of healing is easily done with plants, animals and people.

In the near future, it may be necessary for individuals to take direct steps to solve their own problems, and to take responsibility for self-healing. Many people are already taking constructive action regarding their own health. After all, we're breathing poisonous gas for air, drinking impure water, and eating contaminated food as a result of our present culture. Many people are suffering from mental, emotional, spiritual, and physical disorders and diseases. Unfortunately, most of the disorders come from a social order and environment that we have created ourselves.

All diseases can become curable by improving the environment and by changing our own attitude and lifestyle. The use of an Energy Wand utilizing crystal energy is actually a simple process that allows the balance of the body to be restored. The affected organism heals itself.

The Atlantean Energy Wand is a very simple, easy to build device. It contains the essence of ancient yet advanced science. The energy used is natural and basic to life; its operation is simple.

The basic rod is a hollow copper tube with a copper cap at one end. At the other end is a quartz crystal, about 3/4" to 1" in diameter, and about 1-1/2" to 3" long. The tip of the crystal should have clear, unchipped facets at its point (the six-sided point of the crystal). The outer covering is leather, wrapped in a flat spiral. Some Healing Wands have quartz crystal in both ends, and some have just one crystal with a copper end cap in the other end.

The copper tube acts as an energy accumulator, or funnel; the leather is an insulator, and the quartz crystal acts as an energy transducer, capacitor and focus for the beam of energy. With no moving parts, the only thing that moves through an Energy Wand is energy, and much of its power and efficiency is dependent on its operator. As soon as construction is complete, it begins to radiate energy in all directions from the crystal. When the user focuses it by pointing and thinking/visualizing, a blue-white ray of energy beams in that direction from the point of the quartz crystal at the end. Intensity and distance of the beam are determined by the thoughts, amplified by the emotions, of the individual operator.

These devices are capable of delivering a large amount of accurately focused energy at short or long distances with little or no time delay involved. The user does not have to move or travel to affect its usage. These crystal energy tools, in the hands of an experienced operator, have the ability to reach out through space and time, transforming both energy and matter on a subatomic level.

Transforming radioactive substances into harmless elements is now possible, as is removing other harmful substances from the Earth like poison chemicals, dirty air, contaminated food and water. These transmutations are completely natural processes. The Earth will disperse, transform, and purify itself over a long period of time; these devices merely speed up the process.

The Crystal Wand is in actuality a miniature linear accelerator. It is a 12"-long subatomic particle beam generator. These are just smaller versions of the most powerful machines ever built by man. Many subatomic particles travel at nearly the speed of light. They also travel through all materials: earth, wood, metal, concrete, deep space, etc.

Many engineers and researchers agree that as advances in sophistication are made, machines become simpler instead of more complex. This is obviously the case with crystal devices. They are in balance with the universal laws and physics, being both sensible and practical.

The fundamental application is balance and healing, whether applied to people, plants, animals, poisoned air, polluted foods, contaminated water, or our chemically and radioactively suffering Earth Mother. Total overall balance and healing is a top priority for many of us in this lifetime.

Another fascinating device to make is an Atlantean Crystal Headband. This is a tool to amplify thoughts, and is composed of a copper band with a silver disc and a clear-tipped quartz crystal on top of the silver disc. It focuses and beams these thoughts to specific destinations, depending on the wishes of the operator. The quartz crystal, in conjunction with the copper band and silver disc, acts as a capacitor for storing the energy and a transducer for changing its form. Thoughts and images are converted at the speed of light into beams which are sent to a receiver at some distance or time. The receiver again converts the stream of particles back into images, thoughts and emotions. This is similar to the principles involved in radio and television, except this crystal tool is much simpler.

This device augments psychic impressions, and in many cases, this can be extremely helpful. Be aware that it will amplify *all* emotions and thoughts, even negative ones, so until you become proficient with it, make sure you adopt a very positive attitude when using the Crystal Headband.

Photo by Gary Hulin

The Atlantean Crystal Headband and Power Wand

An effective crystal device is the Crystal Meditation Headband. Sew three circles (3" in diameter) of metallic cloth to a 12" length of moire silk. Epoxy eight rounded, polished crystals in a rosette pattern on the circles. Lay the cloth on top of the head so that the first rosette hangs over the third eye or brow chakra. The middle rosette covers the top, center of the head, and the third rosette hangs at the back of the head—balancing the energy crystals at the brow.

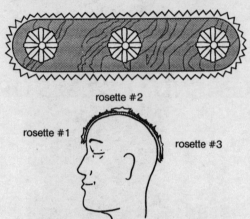

rosettes of polished crystals epoxied on
silver metallic cloth which is sewn to moire silk

rosette #2

rosette #1

rosette #3

Meditation Headpiece

While copper has been used in these dramatic
devices, there is another step upward you may
wish to take in constructing a crystal tool: Use
only sterling silver or coin silver whenever a
metal is needed in combination with quartz crystal, because there is no finer electrical transmission metal than silver—it is the very best, and is
much more efficient than copper, albeit more
expensive.

Use of crystals will become more and more
popular and necessary in the future. If you
haven't already experimented with crystal energy, you are missing out on some valuable assistance from our friends in the mineral kingdom.
You owe it to yourself to become acquainted
with this simple, inexpensive form of energy.

Chapter Nine

TRADITIONAL AND LEGENDARY USES OF CRYSTAL

There are many other gemstones besides clear quartz crystal which can be enjoyable and beneficial. Ageless traditions and legends have endowed these stones with stories of healing and being energized from wearing them, rubbing away trouble spots on the body, or making elixirs with them. Although not having the properties of quartz crystal, many of these attractive gemstones have also been used for healing, meditation, energy, and enjoyment. (Some of the stones listed in this chapter are actually different forms of quartz.)

Gems come in many different colors, sizes, shapes, and varieties. Their material existence is also shown by their weight, hardness, and chemical composition. Their psychic or spiritual influence is shown to us by the inner geometric construction and vibrations—the shape and mathematical relationship of their crystals.

Nearly all gemstones are crystalline in structure. There are only a few which are not. Amber is fossilized tree sap. Coral is created by an underwater creature, and pearls are secretions.

Opal is a member of the quartz group, but lacks a crystal structure, and is still in a comparatively watery state. Gemstones have a symmetry, and it is their relationship to the light passing through them which makes them so valuable to our well-being.

Some gemstones are radiant, and some are receptive. General colors which are radiant are red: ruby, garnet, amethyst, etc. They can help you change your world. They are creative and expansive. The receptive stones are magnetic and help you to absorb what you need. They help manifest what you visualize, and draw your ideas into form. They have a feminine quality. Moonstones are receptive, as are green, black and any dark stones. The softer stones are receptive. A few stones are receptive and radiant at the same time, like topaz.

Doctrine of Signatures

There is an ancient law called the Doctrine of Signatures that explains why certain gemstones may be helpful for some illnesses. Briefly, it states that a certain plant, stone or herb that looks like the disease can actually cure the disease. This is a very ancient and occult principle upon which healing was based, and this was used by alchemists as a basic, guiding principle for healing like with like, in terms of similar inner qualities. It was discussed by Hippocrates, has been handed down in folk lore, and was mentioned in the ancient texts of India called the Vedas.

For example, bloodstone would be good for the blood because it has a reddish color in it that looks like healthy blood. Another example would be the use of Tiger Eye for helping eye problems. Tiger Eye possesses a characteristic called chatoyancy (eye-like), a shiny, reflective property, as does the pupil of the eye. Use of this stone has traditionally been thought to impart better sight to the wearer or user. The same qualities are thought to be true for opal. Its luster is eye-like, too, and because it looks like an eye, the opal has been used to treat eye problems.

The great Renaissance physician and alchemist Paracelsus (who, incidentally, established the role of chemistry in medicine) seems to have been the first to coin the phrase of "like treats like." The principle of similarity Paracelsus used was the Doctrine of Signatures. This principle is used in homeopathic medicine whereby a small, minute dose of the substance needed will stimulate the body to produce the antibodies needed to get well or stay well—much as a smallpox vaccine works. Homeopathic remedies stimulate the vital force in the body.

Gem elixirs, tinctures, and just the proximity of a gemstone or crystal on the body may very well operate on these same premises in healing. There is certainly some validity behind the lasting legends and traditional uses of gems, especially crystals, in healing.

Gemstone Elixirs

The use of gemstone essences, gemstone water, etc., can be traced back to prehistoric times. Early royalty had magic water jugs carved out of quartz crystals because they believed that this would insure the constant flow of precious water forever from the magic jug. Water was as precious then as it is now. Unfortunately, many are slow to realize that pure water is becoming harder to obtain.

Some people believe this legend was related to the lore of the Holy Grail—that it was a carved crystal drinking cup or chalice. Both water and quartz crystal were religious and revered by early people.

Later, gemstones were ground up and fed to royalty, popes, etc. when they were ill, as it was thought that these ground-up potions would help cure them. (What it probably did was hasten their end, as drinking the ground up minerals would just act as sediment in the body, completely indigestible!) Another spinoff of this was the treatment of water with crystal which was given to the sick. Water was left overnight in a container with one or more pieces of crystal in the water (not ground up).

Irish and Scottish farmers were known to soak crystals in a bucket of water and give that special "crystal water" to a sick cow or horse.

Wise women, healers and metaphysicians believed that if the crystals were held in the hand and prayers were said for recovery, this "essence" or vibration placed in the crystal would then in some manner be "charged" in the water, and thus help the ailing animal. This is a sort of formalized ritual for prayer healing and faith healing. It was believed that if the farmer loved the cow and prayed for her earnestly and with faith and love, this would become a thought transmission. The thought would be projected into the crystal, which in turn would put energy into the water. The cow would drink the water and then its body cells and immune system would be greatly activated, thus helping the cow to recover.

Science has recently discovered that interferon in the immune system can be turned on and off by mind control, suggestion, or outside stimulation. Perhaps this too explains why some of these ancient traditions with gemstones actually do work.

What Stones to Use?

There have been basic guidelines throughout history which recommend what stones to use for a given situation, but it is always best to listen to your own intuition. The stone or color you are attracted to at a given time has the vibrations your body needs for energy or balancing. When you begin to listen to your "sixth sense," your intuition, most likely it's your Higher Self speaking to you through inner channels. You may be drawn to a particular stone—attracted to it, desire to have it or wear it. You probably need what that stone has to offer; and the information it has to give to you. Listen to what your body is telling you. You are intuitively sensing the vibration of that particular stone and its effect on your personal energy pattern or being.

There is a great "truth detector" that your body has built into it. If you read, learn, hear, or see something which you believe to be true, something that strikes a chord and rings true with your Higher Self, you will feel chills coursing through your body ("goose bumps"). This feeling is fleeting, and you must be still within to be aware of it. Know that this feeling accompanies a revelation of "truth" for you. If a gemstone elicits this feeling in you, listen to what the stone is telling you.

Jewelry made from gemstones is especially helpful, because then the vibrations from the stone are close to your body and an interaction between the stone and your body cells can take place more easily. For instance, if a person needs to function with some aspect of their emotional understanding, a rose quartz necklace is helpful.

An excellent addition to have in your collection is a lodestone. This is a very inexpensive stone, and is valuable to have because it has the ability to balance the energies of your other stones, and it helps to focus their beneficial aspects. The different stones or crystals can then work together in a more cohesive manner.

Following is a listing of some of the less expensive gemstones readily available, along with descriptions of them, their qualities and some of their uses.

Among other things, they have been used for healing, well-being, energy and good luck. Each stone has its special energies, its unique qualities and vibrations indigenous to its place of origin. They freely share their special gifts and energies with you.

Amazonite

Amazonite is a light aqua-green stone with white mottled flecks. Much of it comes from the New England states and Colorado. It is composed of potassium feldspar, and is a green variety of microline (a potassium aluminum silicate). The name is derived from *amazon stone*, from the Amazon River. While amazonite is found in Brazil, it is *not* found by the Amazon River!

It is a sacred stone highly valued and used extensively by the ancient Egyptians. Amazonite is cooling and soothing to your mental state. This stone is important for healing and spiritual growth. It helps align the heart and solar plexus chakras. It also aligns the etheric and mental bodies. As a thought amplifier, amazonite magnifies the consciousness stored in the upper chakras, especially the psychological attributes.

This pleasing, calming stone makes it easier

for the life force to act as a bonding agent, and this penetrates to the molecular level. It's an enhancer for most other vibrational remedies. On the cellular level, the brain processes are stimulated. All the body energy currents are strengthened by amazonite.

Amethyst

Amethyst is a regal violet gemstone with whitish stripes. The purple color comes from the presence of ions, defects, or inclusions. If ions are the color cause, the color appears after natural irradiation by uranium or thorium.

Quartz crystal is basically colorless. When it is colored, such as amethyst, this is due to either substitute ions, foreign ions, or defects and inclusions in the crystal structure during its formation. It is a form of crystallized quartz, composed of silica. It is found in Brazil, Canada, Africa, Madagascar, the United States, Mexico, Switzerland, China, etc.

Amethyst is a radiant gemstone, meaning that its energies are expansive. It is said that when you meditate with amethyst you are helping the Earth, because the violet ray will help to transform the entire world into a better place. It is the most highly valued stone in the quartz group. Precious (electronic) amethyst is fully transparent. It is purple with red flashes, and costs from six dollars to eighty dollars a carat. It has many supernatural powers. It is said to bring luck, ensure constancy, protect against magic and homesickness. It has long been known to help against drunkenness.

Violet has a calming effect upon the nervous system. Insomnia may be relieved by gently rubbing an amethyst on the temples or forehead, and it can be used for tension and migraine headaches. It is one of the best to use for meditation. It is here to teach the lesson of humility, to "Let go and let God." Amethyst is very useful for people grieving over lost loved ones, as it subliminally communicates that there is no death. Amethyst is recommended for stimulating greater love, and attunement for healing forces. In directing the energy of the amethyst to the lungs, relief may be obtained for asthma and circulation problems. You can recharge your own energy by holding an amethyst over the crown chakra, third eye, or heart chakra. A very high vibration centered in love, balance, and harmony will be transferred.

Apache Tear Drop

Apache Tear Drop is a form of black obsidian. It is a calming translucent stone, found in Arizona and other parts of the U.S. It is composed of feldspar, hornblend, biotite and quartz. It was formed by rhythmic crystallization which produces a separation of light and dark materials into spherical shapes, and is a form of volcanic glass.

There is a haunting legend about the Apache Tear Drop. After the Pineal Apaches had made several raids on a white settlement in Arizona, the military regulars and some volunteers trailed the tracks of the stolen cattle and waited for dawn to attack the Apaches. The Apaches, confident in the safety of their location, were completely surprised and outnumbered in the attack. Nearly fifty of the band of seventy-five Apaches were killed in the first volley of shots. The rest of the tribe retreated to the cliff's edge and chose death by leaping over the edge rather than die at the hands of the white men.

For years afterward, those who ventured up the treacherous face of Big Pacacho in Arizona found skeletons, or could see the bleached bones wedged in the crevices of the side of the cliff.

The Apache squaws and the lovers of those who had died gathered a short distance from the base of the cliff where the sands were white, and for a moon they wept for those who died. They mourned greatly, for they realized that not only had their seventy-five brave Apache warriors died, but with them had died the great fighting spirit of the Pineal Apaches.

Their sadness was so great and their burden of sorrow so sincere that the Great Father imbedded into black stones the tears of the Apache women who mourned their dead. These black obsidian stones, when held to the light, reveal the translucent tear of the Apache. The stones bring good luck to those possessing them. It is said that whoever owns an Apache Tear Drop will never have to cry again, for the Apache maidens have shed their tears in place of yours.

The Apache Tear Drops are also said to balance the emotional nature and protect one from being taken advantage of. It can be carried as an amulet to stimulate success in business endeavors. It is also used to produce clear vision and to increase psychic powers.

Black obsidian is a powerful meditation stone. The purpose of this gemstone is to bring to light that which is hidden from the conscious mind. It dissolves suppressed negative patterns and purifies them. It can create a somewhat radical behavior change as a new positive attitude replaces old, negative, egocentric patterns.

Aventurine

Aventurine is a pleasing dark green stone with a metallic iridescence or spangled appearance (the spangles are small hematite plates). Aventurine is quartz crystal spangled with shiny flakes of mica crystal, and is a very beautiful stone. It is found in India, China and Brazil.

It is said to bring luck and adventures in love and games. It makes an individual independent and original. It has a binding and healing force, and is good for skin diseases and improving the complexion. At one time it was used to cure nearsightedness. It is helpful for the etheric, emotional and mental bodies. Aventurine has strong healing energies, and affects the pituitary gland. It can be used for creative visualization, Higher Self attunement, and is good for the muscle and nervous system.

Aventurine is a suitable stone for artists, writers and all those of a creative nature. It helps bring prosperity; the green vibrations attract money.

Bloodstone

Bloodstone is a green opaque stone with spotted red flecks. It is a member of the cryptocrystalline quartz group, and is composed of silica. Bloodstone belongs to the general group of chalcedony and is found in India, Australia, Brazil, China and the U.S.

Bloodstone contains deep earth-green and a deep, blood red. Together these create a powerful cleanser for the physical body. It is an important purifier for the kidneys, liver, spleen and blood. In times past, bloodstone was used to stop bleeding and hemorrhage by wounded soldiers and mothers-to-be. It will detoxify the body. Bloodstone helps in transforming the body to be able to carry more light and energy.

Legend has it that when Christ was crucified, the blood from his wound dripped to the green jasper ground, spotting it red and thus forming this stone. This stone was also known as heliotrope, and it was believed that if one covered the stone with the herb heliotrope, the owner became invisible. This combination was also used in many other magical rites.

Blue Lace Agate

Blue Lace Agate is a beautiful, pale blue

stone with concentric marking. It is a cryptocrystalline quartz stone. Agate is banded chalcedony, the bands having been formed by rhythmic crystallization. Agates are found as nodules or geodes in siliceous volcanic rocks. This stone comes from southwest Africa.

Agate strengthens the power of the Sun in your astrological sign when you wear it. It helps you stay well-balanced. It sharpens the sight, illuminates the mind and helps you speak.

Blue lace agate helps you develop and realize your inner peace. These stones affect the physical body, first at the densest levels, and then at the levels of some of the higher bodies as well.

Carnelian

Carnelian is a translucent orange-red stone. It is cryptocrystalline quartz, composed of silica. It is found in India and South America.

In ancient times carnelian was thought to still the blood and soften anger. It is a gem of the Earth, a symbol of the strength and beauty of our planet. It is good for people who are absent-minded, confused or unfocused. It strengthens the voice and helps one become aware of the subtle body (the invisible, ethereal substance which interpenetrates the physical body and extends outward about five to eight inches. It is invisible because it vibrates faster than the five physical senses can perceive). Carnelian gives you a feeling of consolation despite the hardships of life. It has been considered symbolic of the third eye, and is the symbol of the spiritual love of good. It helps to banish fear. It is a good, general healing stone.

Jade

Jade is a soothing green color. It is an avocado green gem, with darker mottled flecks of green in it. Although Jade is commonly thought of as being green, it can be found in most colors from white to black, and in between. Some of the better varieties are almost transparent. The green variety of Jade comes from Wyoming. It is composed of sodium aluminum silicate, and because of its felt-like structure, it is very tough and resistant.

The name goes back to the time of the Spanish conquest of Central and South America and means *piedra de ijada* (hip stone), as it was used as a protection against and cure for kidney diseases. Jade is the prince of peace and tranquillity. It acts quietly as a consciousness raiser of human development. It dispels negativity by the constant emission of soothing and cleaning vibrations.

Moonstone

Moonstone is a translucent gem which is either a light toned gem of many hues, or it can be colorless, with a milky-blue sheen. It is the most important gemstone of the feldspar group. It is composed of potassium feldspar. (Feldspars are silicates of aluminum and either potassium, sodium or calcium.) It comes from Ceylon or Brazil.

Moonstone is a receptive stone. It helps you to balance and soothe your emotions so that you don't have to react from an emotional state. It helps your Higher Self control your emotions so you can grow more spiritually. Moonstones help you to experience calmness and peace of mind.

These stones help women's hormonal and emotional equilibrium, and they help men become more in tune with the feminine side of themselves. The Moonstone can act as a magical link so that your guides can communicate with you easier in order to show you what your life path really is. The moonstone can hold charges in it, and may need to be cleansed occasionally.

Opal

Opal is a noncrystalline form of quartz—a silica gem which contains varying amounts of water (3 to 9 per cent). Technically, opal is a hydrous silica existing in a colloidal state. Opal forms as a low temperature deposit around hot springs and veins. Opal is found in Nevada. The common opal is milky white, but it can be found in greenish yellow to brick red. A beautiful form of opal known as fire opal is found in Mexico. It has luminescent reds and purples in it. Australian opal deposits are noted worldwide.

Opal is one of the most mysterious of all gems. Some legends state that opal is bad luck for all those other than whose birthday it rules (October). Others believe that it is all right for anyone to receive an opal as long as it is given as an act of friendship. The opal is said to help in all diseases of the eyes, and to sharpen and strengthen the sight.

Petrified Wood

Petrified wood, also known as fossilized wood, is a gray-brown conglomerate of muted tones. It can have light brown, yellow, red, pink, and even blue to violet colors in it. It is a microcrystalline quartz and a member of the chalcedony family. The organic wood is not really changed into stone, only the shape and structural elements of the wood are preserved. It is found mainly in the southwest U.S.

Petrified wood is very earthy, and will assist you in becoming grounded and balanced. If you feel spacey and not quite "with it," having a piece of petrified wood near you will restore your subtle bodies to a more harmonious grounded state, and you will be able to think and reason more clearly.

It was used by the American Indians as a protective amulet against accidents, injuries and infections. It was thought to bring good luck, build reserves of physical energy, help ease mental and emotional stress, and encourage emotional security.

This stone is helpful for arthritis, environmental pollutants, and skeletal systems; it enhances longevity and is a general strengthening help for the body.

Rhodonite

Rhodonite is an attractive rose pink stone with black veinings. It is composed of manganese metasilicate. The name comes from its color (Greek for rose). It is found in Canada, the U.S., and Mexico.

Rhodonite is good for mental unrest and confusion, anxious forebodings and incoherence. It fends off unwanted influences from the etheric planes. It is good for psychically sensitive people who would prefer to be left in peace. It relieves anxiety and stress, promotes mental balance and mental clarity. Rhodonite helps one deal with sensitivity, self-esteem, and also helps one become more self-confident. On the physical level, it is good for the skeletal system.

Rose Quartz

Rose Quartz is a gentle, pale to medium pink, translucent form of quartz. It comes from Brazil. This stone has a soft and useful frequency, and does not conflict with any other stones. The best form of rose quartz is transparent.

It is an important stone for the heart chakra, and for giving and receiving love. It helps to dissolve all burdens and traumas that have burdened the heart. Rose quartz assists in understanding and dissolving problems so that the heart is better able to know love. As its presence is felt by the body, sorrows, fears, and resentments are replaced by a sense of personal fulfillment and peace.

This stone promotes the vibrations of universal love and inner serenity. It teaches that the many negative childhood experiences each one undergoes enable the self to learn how to love and nurture itself. It also enlivens the imagination, enabling one to create beautiful forms.

Rutilated Quartz

Rutilated quartz is clear or smoky quartz with threads of titanium dioxide (gold or silver filaments) running through it. Much of this stone comes from Brazil. It is also called needle stone or angel's hair.

Rutilated quartz energizes, rejuvenates, and balances the system. It raises your vibrations, increases clairvoyance, and strengthens thought projections. This gemstone helps the body in the assimilation of nutrients, helps the immune system function more effectively, slows diseases of aging, and prevents depression.

The crossing of the rutiles in this type of quartz represents the accord of tissue regeneration within the physical body. This mineral also stimulates the electrical properties of the body.

Smoky Quartz

Smoky quartz is a translucent grayish/brown quartz with natural irradiation from the Earth. The colors vary from light smoke to almost black. The best quality is fully transparent. Much of it comes from Brazil.

This form of quartz will initiate movement of the basic, primal forces of your body, allowing you to better express your physical self. Smoky quartz will lend a person a sense of pride to be able to walk the Earth and inhabit a human body. It is very helpful for those of an earthy nature, and when worn as an amulet it can induce mental clarity and stimulate physical energy. It also protects and strengthens one while walking on the Earth. Smoky quartz is a very grounding stone, and if you are feeling spacey and unbalanced, wearing it or holding it will give you a connection with the Earth.

Snowflake Obsidian

Snowflake obsidian is a striking black, lustrous opaque stone with grayish/white bold markings, much like the beautiful patterns of snowflakes on a black background. It is a form of volcanic, amorphous, siliceous glassy rock. This form of obsidian is found in Utah.

Snowflake obsidian is said to sharpen both the external and the internal vision. It is one of the most important "teachers" of the New Age stones. It is the warrior of truth, and shows the self where the ego is at, and what it must change in order to advance to the next step of evolutionary growth.

With black on one end of the color scale and white on the other, we are shown the contrast of life: day and night, darkness and light, good and evil. The black symbolizes mastery over the physical plane, and the white symbolizes the purity inherent in each one of us. The snowflake obsidian will help you to clear out all the cobwebs in the corners of your mind.

Snow Quartz

Snow quartz is a delicate, translucent white form of quartz. Although it is very attractive and pure looking due to its whiteness, it is not electronic. It is a member of the chalcedony family, and is made of silica. It is a cryptocrystalline quartz and is found in Brazil, the U.S. and Mexico.

Snow quartz helps us to have a focus of purity in ourselves. It promotes clarity of mind, and activates the crown chakra. It shows us our personal identification with the Infinite, the oneness with God. It represents peace and wisdom.

It is a stone that has the power to act as an insulator for all things. It can stop negative vibrations. Snow quartz can help develop psychic abilities. It causes the intellect to become more spiritual and helps one to have a love of truth.

Sodalite

Sodalite is a deep, rich blue stone with white inclusions. It is composed of chloric sodium aluminum silicate. It is found mostly in Canada and Brazil.

Sodalite is said to prolong physical endurance, and is used by athletes. It is said to help create harmony within the inner being and to stop conflict between the conscious mind and the subconscious. It is good for those who are oversensitive and reactive, allowing a person to shift from emotional to rational thinking. Sodalite helps clear away old mental patterns.

It helps one to understand the nature of one's self in relation to the universe. It awakens the Third Eye, which prepares the mind to receive the inner light and intuitive knowledge. Sodalite is the densest and most grounded of the deep blue stones, and clears the mind so that it can think with greater perception.

Tiger Eye

Tiger eye is a beautiful, golden-brown shiny stone which appears lifelike due to its chatoyancy, or silky luster. It is crystallized quartz, made of silica. It is found mostly in Africa.

This stone has been worn through the ages to avert the evil eye and help prevent eye diseases. Tiger eye helps people gain insight into their own faults, and to think more clearly. This stone is helpful for greater spiritual understanding. It helps one develop courage and inner strength, and gives one a sense of responsibility.

Tiger eye helps to defeat negative forces. Because of its ever-changing appearance when viewed from different angles, it helps the person using it to become "all seeing"; able to view different ways of observing a situation. It gives the wearer the ability to become more direct, more channeled in their way of thinking.

Tourmalinated Quartz

Tourmalinated quartz is a clear form of quartz with black, brown, green or silver colored threads of tourmaline running through it. Much of this stone is found in Brazil.

Tourmalinated quartz is good for dissolving fear in oneself. It can aid in eliminating negative conditioning patterns we have experienced in our lives. Use of this type of quartz kept near the body and meditated upon is said to increase mental awareness and enhance psychic ability. It soothes the central nervous system, and helps to alleviate depression and nervous exhaustion. Because it is a combination of quartz and tourmaline, it has influences and characteristics of both these gems.

Unikite

Unikite is a combination of salmon pink feldspar and green epidote, and is an opaque stone. It is named after the Unikite Mountains between North Carolina and Tennessee, where it is found.

The pink in unikite speaks to the heart chakra to awaken the love within. The green lends it healing qualities to any hurts which have been sustained. The pink in Unikite is a deeper shade of pink and is more grounding than that found in rose quartz. This stone has a leveling effect, and helps to balance the emotional aspects of the body. It is an earthy, peaceful stone.

Conclusion

The use of *any* form of quartz crystal will benefit you. What can you do with these gemstones and crystals to help yourself? Meditate with them. Place them on a part of your body that needs healing. Carry them around in your pocket. Wear them as jewelry or use them as a keychain. Sleep with them under your pillow. Put them on your desk, kitchen table, coffee table, anywhere that you will be near them and enjoy the benefits of their vibrations. They are just like a good friend—one whose companionship you delight in. Love them. Look at them. Enjoy them. They help you by their very presence. It's a gift they freely and lovingly give to you.

Fine electronic crystal is Nature's most valuable tool for mind development and control, and is a help we should not be without.

Bibliography

David, William. *The Harmonics of Sound, Color and Vibration,* Marina del Rey, California: DeVorss and Co., 1984.

Gurudas. *Gem Elixirs and Vibrational Healing, Vol. 1,* Boulder, Colorado: Cassandra Press, 1985.

Leadbeater, C.W. *The Chakras,* Wheaton, Illinois: Theosophical Publishing House, 1927.

Raphaell, Katrina. *Crystal Enlightenment,* New York: Aurora Press, 1984.

Sherwood, Keith. *The Art of Spiritual Healing,* St. Paul, Minnesota: Llewellyn Publications, 1985.

Smith, Michael. *Crystal Power,* St. Paul, Minnesota, Llewellyn Publications, 1985.

Stein, Diane. *The Women's Book of Healing,* St. Paul, Minnesota, Llewellyn Publications, 1986.

On the following pages you will find listed, with their current prices, some of the books now available on related subjects. Your book dealer stocks most of these and will stock new titles in the Llewellyn series as they become available. We urge your patronage.

TO GET A FREE CATALOG

To obtain our full catalog, you are invited to write (see address below) for our bi-monthly news magazine/catalog, *Llewellyn's New Worlds of Mind and Spirit*. A sample copy is free, and will continue coming to you at no cost as long as you are an active mail customer. Or you may subscribe for just $10 in the USA and Canada ($20 overseas, first class mail). Many bookstores also have *New Worlds* available to their customers. Ask for it.

TO ORDER BOOKS AND TAPES

If your book store does not carry the titles described on the following pages, you may order them directly from Llewellyn by sending the full price in U.S. funds, plus postage and handling (see below).

Credit card orders: VISA, MasterCard, American Express are accepted. Call us toll-free within the United States and Canada at 1-800-THE-MOON.

Postage and Handling: Include $4 postage and handling for orders $15 and under; $5 for orders *over* $15. There are no postage and handling charges for orders over $100. Postage and handling rates are subject to change. We ship UPS whenever possible within the continental United States; delivery is guaranteed. Please provide your street address as UPS does not deliver to P.O. boxes. Orders shipped to Alaska, Hawaii, Canada, Mexico and Puerto Rico will be sent via first class mail. Allow 4-6 weeks for delivery. **International orders:** Airmail – add retail price of each book and $5 for each non-book item (audiotapes, etc.); Surface mail add $1 per item.

Minnesota residents please add 7% sales tax.

Llewellyn Worldwide
P.O. Box 64383 L246, St. Paul, MN 55164-0383, U.S.A.

For customer service, call (612) 291-1970.

Prices subject to change without notice.

CRYSTAL AWARENESS
by Catherine Bowman

Crystals have long been waiting for people to discover their wonderful powers. Today they are used in watches, computer chips and communication devices. A spiritual, holistic aspectare also involved with crystals.

Crystal Awareness will teach you everything you need to know about crystals to begin working with them. It will also help those who have been working with them to complete their knowledge. *Crystal Awareness* is destined to be *the* guide of choice for people who are beginning their investigation of crystals.

0-87542-058-3, 224 pgs., mass market, illus. $3.95

CRYSTAL POWER
by Michael G. Smith

This is an amazing book, for what it claims to present—with instructions so you can work them—is the master technology of ancient Atlantis: psionic devices made from common quartz crystals! Learn to construct an "Atlantean" Power Rod, Crystal Headband, or Time and Space Communications Generator.

These crystal devices seem to work only with the disciplined mind power of a human operator, yet their very construction seems to start a process of growth and development, a new evolutionary step in the human psyche that bridges mind and matter.

Does this "re-discovery" mean that we are living, now, in the New Atlantis? Have these Power Tools been re-invented to meet the needs of this prophetic time? Are Psionic Machines the culminating Power To the People to free us from economic dependence on fossil fuels and smokestack industry? This book answers "yes" to all these questions, and asks you to simply build these devices and put them to work to help bring it all about.

0-87542-725-1, 256 pgs., 5 1/4 x 8, illus., softcover $9.95

CUNNINGHAM'S ENCYCLOPEDIA OF CRYSTAL, GEM & METAL MAGIC
by Scott Cunningham

Here you will find the most complete information anywhere on the magical qualities of more than 100 crystals and gemstones as well as several metals. The information for each crystal, gem or metal includes: its related energy, planetary rulership, magical element, deities, Tarot Card, and the magical powers that each is believed to possess. Also included is a complete description of their uses for magical purposes. The classic on the subject.

0-87542-126-1, 240 pgs., 6 x 9, illus., softcover $14.95

CRYSTAL VISION
Shamanic Tools for Change & Awakening
by Michael G. Smith & Lin Westhorp

Crystal Vision gives you directions on how to perform your own miracles through advanced crystal technology that taps directly into powerful Universal Energy. No matter what your level of technical expertise, this book will enrich your experience of and participation in this exciting era of human evolution. Use crystals where they'll have the most impact on your daily life. Explore the realm of high-tech black box psionics with the "Psi-Comp," an astonishingly powerful 21st Century tool that uses modern technology to help you boost the power of your mind so you can take more control of your life.

Explore tools made from archetypes that will help you evolve into the consciousness of your spiritual essence. Through making and using these tools: pyramids, the Ankh Crystallos, Trident Krystallos, Crux Crystallum and Atlantean Crystal Cross, you can develop your inborn talents to create the world of your dreams. What's more, you will discover some of the most powerful shamanic crystal tools in use today.

0-87542-728-6, 272 pgs., 6 x 9, illus.,   softcover$12.00

Prices subject to change without notice.